Discrete
Iterated
Function
Systems

Discrete Iterated Function Systems

Mario Peruggia

Department of Statistics
The Ohio State University
Columbus, Ohio

A K Peters
Wellesley, Massachusetts

Editorial, Sales, and Customer Service Office

A K Peters, Ltd.
289 Linden Street
Wellesley, MA 02181

Library of Congress Cataloging-in-Publication Data

Peruggia, Mario.
      Discrete iterated function systems / Mario Peruggia.
         p. cm.
      Includes bibliographic references and index.
      ISBN 1-56881-015-6
      1. Image processing—Digital techniques—Mathematics.
2. Fractals. I. Title.
TA1637.P47 1994
621.36'7—dc20   93-20664
                                              CIP

Printed in the United States of America
97 96 95 94 93   10 9 8 7 6 5 4 3 2 1

In memory of my mother Anna Maria

Contents

Preface

A novel method for encoding a large class of digitized color images has enjoyed wide popularity ever since Michael Barnsley's book *Fractals Everywhere* was published in 1988. The method is based on the identification of an image with the unique stationary distribution of an ergodic Markov chain, which has the Euclidean plane as its state space.

The probability structure of the chain is given in terms of an iterated function system (IFS) with probabilities (a finite collection of contractions of the plane into itself, and associated weights). Decoding of an image is obtained by plotting the points along a trajectory of the Markov chain as they are generated; color is added by means of some prespecified correspondence between frequencies and color intensities.

When each transformation comprising an IFS is approximated by a discretized version that takes on values in a grid constituting a partition of the plane, we obtain what may be called a *discrete* IFS. In this book, we examine the theoretical and practical consequences that are brought about by this discretization. In particular, we explore the connections between continuous and discrete IFSs, point out the differences, and illustrate some surprising implications arising from the discretization procedure.

The book should be accessible, for the most part, to readers who have had some background exposure to elementary notions about topological and metric spaces, probability theory, and Markov chain theory. While certain technical details may require a slightly higher degree of mathematical sophistication to be fully understood, the explanations and examples we provide should always suffice to convey effectively the main ideas.

Chapter 1 contains a brief introduction, while the background theory is reviewed in Chapter 2 and in the first six sections of Chapter 3. Specifically, Chapter 2 deals with the deterministic approach to the theoretical development of IFS theory, while the probabilistic approach is outlined in Chapter 3. This review is intended to provide all the information about IFS theory needed in the sequel. In the final section of Chapter 3, we present some interesting theoretical results not directly related to the problem of representing digitized images. In particular, we provide a novel characterization of all Beta distributions with integer parameters in terms of IFSs with probabilities.

The theory of discrete IFSs is fully developed in Chapter 4, where, starting from a continuous IFS, we introduce a Markov chain, called the *round-off process,* whose state space consists of a *discrete* grid over the plane. In studying this new process, we were originally motivated by the desire to construct an approximate model for the propagation of rounding errors caused by the implementation of the image generation algorithm based on the original chain in a computer program.

We first show that the round-off process corresponding to a given IFS may have more than one stationary distribution, and yet, as the grid spacing decreases, all these distributions converge weakly to a single limit, given by the unique stationary distribution of the original process. We then use these results to discuss the impact of rounding errors on the implementation of the original image generation algorithm.

The definition of the discrete chain is based on the notion of a *round-off version* of a contraction on the Euclidean plane. In studying its behavior, we discover that its domains and subdomains of attraction present varied and unusual shapes, of which we display several examples. The variety and complexity of these shapes is all the more surprising in view of the simple nature of the defining transformation and discretization procedure.

Finally, in Chapter 5, we review how IFS theory can be applied to produce animated motion pictures, and discuss some of our experiences in the field. Several illustrative figures are presented throughout the book.

The book constitutes a revised version of the author's Ph.D. dissertation, completed in 1990. As a consequence, most of the referenced material predates it. We are convinced that the main sources we consulted when writing the original manuscript—in particular, the seminal paper by John Hutchinson, and the writings by Michael Barnsley, Marc Berger, and their coauthors—still provide excellent and up-to-date accounts of IFS theory. They also supply all the background information needed for the development of the original material contained in this book.

Owing to the vast popularity of the subject, numerous publications, both of a technical and expository nature, have appeared in the intervening time. While we have made no consistent attempt to update the bibliography, we have, however, added a few new references (cited either within the text or in footnotes) to complement and integrate those contained in the original manuscript, and to indicate some of the new directions in which research in this field is currently headed.

Acknowledgments

This book is largely based on the Ph.D. dissertation (Peruggia (1990)) that I wrote while studying statistics at Carnegie Mellon University in Pittsburgh, Pennsylvania. There are several people to whom I am grateful for the completion of this work. First, I would like to thank my adviser, William Eddy, not only for his apt and thorough supervision of my thesis research, but also for his helpful guidance throughout my whole stay at Carnegie Mellon.

I am also immensely indebted to Marc Berger, who first introduced me to the subject of iterated function systems, and whose insightful comments and suggestions have always proved invaluable. The final version of this work has also significantly benefited from the criticism of the other two members of my reading committee, David Banks and Mark Schervish, and from the careful proofreading of Eric Vognild.

Of great help in organizing my ideas and improving my knowledge was the exciting course on fractal geometry taught by Steven Shreve in the fall of 1989. My research makes also use of the theory of stochastic processes that I learned from John Lehoczky. I thank him for fostering my interest in the subject with his stimulating lectures.

I would like to thank Alice and Klaus Peters who gave me the opportunity to bring my work to the public, and the editorial staff at AK Peters for their technical support with the preparation of the final version of the manuscript.

The manuscript was typeset using the Latex document preparation system. The illustrations were prepared with the Drawcgm and Gplot software packages developed at the Pittsburgh Supercomputing Center. The *xv* software package was also employed for the final editing of the cover image and of the color insert.

I am certain that completion of this work would have been much more difficult without the encouragement and moral support of my family and my friends. To the memory of my mother, who died on May 17, 1992, to her inspiring example of industriousness and rectitude, I would like to dedicate this book.

Finally, I would like to acknowledge that this research was partially supported by the Consiglio Nazionale delle Ricerche of Italy under Fellowship No. 203.01.36, and by a 1991 Seed Grant from The Ohio State University.

Columbus, Ohio, 1993   *Mario Peruggia*

List of Figures

1

Introduction

The importance of *data compression* is constantly growing, as the need for storage and transmission of large volumes of data increases. The basic idea is to reduce the number of bits required to represent stored or communicated data, by means of encoding methods aimed at eliminating redundancy in the data, while preserving its original information content. Typically, a data compression method consists of an encoding scheme, through which a string of characters in some representation is transformed into a new, shorter string, and a decoding procedure that reverses the encoding process and allows recovery of the original representation of the string. The goodness of a data compression technique is to be judged both on the basis of the attainable levels of compression and of the time required by the encoding and decoding processes.

A careful exposition and development of these ideas can be found in Lelewer and Hirschberg (1987). The authors there present an extensive review of methods for data compression, distinguishing between *general-purpose* and *semantic-dependent* techniques. General-purpose techniques are characterized by the fact that they can be employed without any knowledge about the information content of the data. Semantic-dependent techniques, on the other hand, achieve compression by exploiting the particular nature of the data in order to reduce redundancy.

Data compression is extremely important in the area of image representation and processing. Digitized images are usually stored as large arrays, in which each entry contains information about the color and intensity of a corresponding pixel on the screen. This kind of representation, however, is very expensive. For instance, in the common situation where there are 256 brightness levels, 7,077,888 bits are needed in order to capture an image of 1,024×864 pixels. [1] Good image data compression can be achieved through semantic-dependent techniques that exploit the tendency of pixels of similar color and intensity to cluster together (see Lelewer and Hirschberg (1987) for further details).

The method of image generation presented in this book yields enormous data compression for special classes of images. In its simplest version, it is based on the identification of a given black-and-white image with the unique fixed point of a contractive transformation on the space of nonempty, compact subsets of the Euclidean plane, endowed with the Hausdorff metric. Such a transformation is defined in terms of a finite collection of contractive mappings of the Euclidean plane into itself, which is often referred to, in the literature, as a hyperbolic *iterated function system* (IFS), while the fixed point of the transformation is usually called the *attractor* of the IFS.

The theory was first presented in a general framework by Hutchinson (1981). The author there considers IFSs on a generic, complete metric space, and thoroughly examines the self-similarity property of their attractors. Self-similarity is indeed a characterizing property of these sets, which, on the one hand, defines and limits the class of images that are amenable to being encoded through IFSs, and, on the other, constitutes the very source of data compression.

From a technical point of view, self-similarity means that the attractor of a hyperbolic IFS can be seen as the finite union of its images under the mappings that constitute the IFS. In practice, it tells us that a given image, when regarded as a compact subset of $\mathbb{R}^2$, can be encoded through an IFS if there exist finitely many contrac-

---

[1]Size of a VAXstation VR290 color monitor.

tive transformations on the Euclidean plane such that, by applying the transformations to the image and taking the union of the sets so obtained, the original image is recovered. Since the transformations involved need not be similitudes, this property is perhaps better referred to as self-covering. The latter terminology is also widely used in the literature, and we shall adopt it throughout the remainder of this book.

The case when all transformations employed are affine, i.e., the case when the transformations consist of a linear part plus a shift, is extremely important in practice. This is so since the image of a compact subset of the plane under an invertible affine transformation is a compact subset of the plane whose shape still shows resemblance with that of the original subset. Because of the self-covering property then, the attractor of a hyperbolic IFS on the plane can be viewed as a collage of affinely similar, and possibly overlapping, pieces. On the other hand, if an image is given which appears to be the union of a finite number of affine copies of itself, then, by determining which transformations correspond to the various elements of the union, one obtains an IFS that fully describes the image. Since an affine transformation on the plane is defined in terms of six real parameters, it is clear that being able to represent an image as the attractor of an IFS consisting of a small number of affine transformations constitutes a valuable source of data compression.

The development of IFS theory in a deterministic setting is paralleled by a probabilistic development that allows consideration of full-color images, and gives rise to an extremely effective algorithm for image generation. Early references on the subject are the already mentioned article by Hutchinson (1981) and the article by Barnsley and Demko (1985). A few more recent references are, in chronological order, Diaconis and Shahshahani (1986), Barnsley et al. (1986), Berger and Amit (1987), Elton (1987), Barnsley et al. (1988), Berger (1988), Berger and Soner (1988), and Berger (1989b). While all the previous references are rather technical, some nontechnical survey articles are also available. Among those we can mention are Demko et al. (1985), Barnsley and Sloan (1988), Zorpette (1988), and Berger (1989a). Finally, we cannot fail to mention the beautiful book by Barnsley (1988), which contains a thorough exposition of the theory and a wealth of exciting images, and the more recent book by

Falconer (1990).

The basic idea behind the stochastic approach consists in assigning a positive probability to every transformation in a given hyperbolic IFS, thereby constructing what is known as an IFS *with probabilities,* and defining a Markov chain on the plane, whose transition probability structure is specified in terms of the transformations in the IFS and the associated probabilities. Such a Markov chain has a unique, invariant, limiting distribution, whose support coincides with the attractor of the IFS, when the IFS is regarded as deterministic. In addition, for any starting point, the empirical distributions of almost every orbit of the chain converge weakly to its unique, invariant distribution.

These results have immediate application to image encoding and decoding. Encoding is accomplished by identifying a given image with the invariant distribution for an appropriate IFS with probabilities on the plane. The support of the invariant distribution determines the shape of the image, while the additional probabilistic structure is employed to represent color, by means of some prespecified correspondence between frequencies and color intensities.

The convergence result about the empirical distributions of orbits of the Markov chain is the basis for a stochastic generation algorithm. Decoding of the image is obtained by running a single orbit of the Markov chain associated with the IFS that provides the encoding, and plotting the points in the orbit as they are generated. Since the empirical distributions of almost every orbit converge weakly to the unique, invariant distribution for the IFS, one is guaranteed that the resulting image will always look the same, regardless of the particular orbit that has been generated. When all transformations involved are affine, the computational effort required to generate an orbit of the Markov chain is minimal, and the stochastic algorithm provides an extremely efficient decoding procedure, which is also well-suited for parallel implementation.

Extensions and generalizations of the basic probabilistic approach are presented in several of the above-mentioned references. One possible way in which the assumptions can be relaxed is by requiring that not all transformations in the IFS be contractive, but that they only be contractive "on the average." The probabilities associated with the individual transformations can also be assumed to be dependent on the current state of the Markov chain, instead of being constant. Another interesting and useful generalization is given by a *mixing algorithm,* which allows one to generate more elaborate images by mixing together, according to a probabilistic scheme, orbits of the Markov chains associated with different IFSs with probabilities.

Chapters 2 and 3 contain mainly background material, which serves the purpose of reviewing the current state of the theory and of making this book self-contained. The deterministic development of IFS theory is outlined in Chapter 2. There, we briefly review the definitions and basic properties of contractive and affine transformations, and provide a rigorous definition of IFSs, together with a discussion of their most significant properties. We then present several examples of IFSs, whose attractors are illustrated in the several figures throughout the chapter. The notion of IFSs with condensation is also introduced, and a possible approach to the solution of the encoding problem, i.e., the problem of determining an IFS representation for a given image, is reviewed. Such an approach, described in several of the above-mentioned references, tends to exploit the self-covering property in order to determine what transformations would yield an IFS whose attractor constitutes a good rendering of the given image.

Chapter 3 deals with the probabilistic approach to developing IFS theory. In it, we describe the basic image generation algorithm, also known as the *random iteration algorithm,* and present the stochastic model upon which it is based. As we have mentioned before, several variations of the basic model have been considered in the literature, including a mixed probabilistic model. Our review is far from being complete, but should be sufficient to provide a clear understanding of the basic ideas, and a statement of the results that we shall use in the sequel. Among other things, we review some of the methods that have been proposed to address the image encoding problem in the probabilistic setting.

The last section of the chapter is the only part of the book that is not directly motivated by the application of IFS theory to image encoding and decoding. In it, we present some results that we have obtained by applying the invariance condition for the invariant distribution of an IFS with probabilities to the problem of computing the integer moments of several random variables. We also show that any Beta distribution with integer parameters can be obtained as a marginal of the invariant distribution for an appropriate IFS with probabilities, on a Euclidean space of suitable dimensionality.

The main results are presented in Chapter 4. Whenever the random iteration algorithm is implemented on a computer to produce a full-color image on a graphics terminal, an orbit of the Markov chain associated with a given hyperbolic IFS with probabilities is generated and its points are plotted. In doing so, the continuous state space $\mathbb{R}^2$ of the Markov chain is approximated twice by discrete grids. The first approximation is a result of the finite precision in which the calculations are carried out; the second is induced by the finite number of pixels on the screen.

The question arises naturally of whether a computed orbit provides a good approximation to the theoretical orbit. We observe first that, in the absence of rounding errors, the discreteness of the screen would have a negligible effect, since the pixel that would be turned on at any given time would actually contain the point visited by the exact orbit at that time. On the other hand, the approximation introduced by floating-point arithmetic could have severe negative effects, if rounding errors had a tendency to propagate.

It is therefore the latter approximation that we proceed to model, by defining a new Markov chain that we call the *round-off process,* whose discrete state space is given by the square elements of a regular grid laid over $\mathbb{R}^2$. This is actually a simplifying assumption, since the computational grid is not regular; floating-point representation implies that the distance between adjacent point numbers depends on the exponent (so numbers large in absolute magnitude have distant neighbors). Nevertheless, we feel that this model gives a fairly realistic description of the propagation of rounding errors. Furthermore, the image generation algorithm that results from applying the random iteration algorithm to the round-off process can be implemented exactly on a computer.

In Section 4.1, we give the formal definition of the round-off process associated with a given hyperbolic IFS with probabilities, based on the notion, which we introduce, of the round-off version of a transformation on the plane. The round-off version of a given transformation is defined in such a way that, when it is applied to a point in the plane, it maps such a point into the center of the element of the grid to which the image of the point under the original transformation belongs. The properties of the round-off version of a single contractive transformation on the Euclidean plane are examined in Section 4.2, and several examples are provided in Section 4.3.

The final section of the chapter deals with the limiting properties of the round-off process, as the size of the elements of the grid, also called the level of accuracy, goes to zero. In particular, it is shown that, for any given level of accuracy, the round-off process has a finite number of stationary distributions, and that, as the level of accuracy goes to zero, each of them constitutes a better and better approximation (in the sense of weak convergence) to the unique invariant distribution for the original Markov chain associated with the given IFS.

In terms of our application to image generation, this result guarantees that the effect of computer rounding does not build up, and that the image generated on the monitor always provides a good approximation to the theoretical image. Since the round-off process might have, at a given level of accuracy, more than one stationary distribution, which image one actually obtains depends, in general, on the starting point of the orbit generated by the random iteration algorithm. Two uniqueness conditions are presented at the end of the chapter.

The final chapter describes how IFS theory can be applied to produce computer animation. This application is based on a continuity result outlined in Section 5.1, and whose proof can be found in Barnsley (1988), which guarantees that, given a family of hyperbolic IFSs suitably parameterized, the attractor is a smooth function of the parameter. Our experiences in this field are summarized in Section 5.3, and the animation equipment that we have employed is described in Section 5.4. Some illustrative figures, representing frames of various animation segments, are presented throughout the chapter.

2

Deterministic Theory of Iterated Function Systems

In this chapter, we intend to present and develop the theory of iterated function systems from a completely deterministic point of view, along the lines followed by Barnsley (1988). It is our intention to focus exposition on those topics and results that are of fundamental importance for understanding how the theory can be applied to the practical purpose of encoding and decoding digitized images. Since such images can be regarded as compact subsets of Euclidean spaces (most notably $\mathbb{R}^2$), we shall always try to provide examples that take place in such spaces, despite the fact that the results to which those examples will be referring hold, in general, in any complete metric space. Emphasis will be placed on those aspects of the theory that we shall need in the sequel.

The purpose of this chapter is therefore twofold: to provide a quick review of the theory of iterated function systems, focusing on examples pertinent to the realm of image generation, and to outline results that will be used in the following chapters, so as to make this document as self-contained as possible. As a result, our treatment will be far from exhaustive, some important topics being completely omitted, and others being only briefly mentioned. Once again, we refer the interested reader to Barnsley (1988) for a more comprehensive and in depth discussion. [1]

In the remainder of the chapter, after mentioning a few elementary facts about metric spaces, contractive transformations, and affine transformations, we review the definitions and fundamental properties of an IFS, and of an IFS with condensation. Throughout, we elucidate the concepts with several examples and illustrations. We conclude the chapter by reviewing how the self-covering property for IFSs can be exploited to find a solution to the image encoding problem.

2.1. Preliminaries

As previously mentioned, the deterministic theory of iterated function systems can be developed in any complete metric space. For any given set X and metric $d : X \times X \mapsto [0, \infty)$, we shall denote by (X, d) the metric space consisting of the set X endowed with the metric d. The metric space (X, d) will often be assumed to be complete, in the sense that every Cauchy sequence of points in X is required to converge to a limit in X. Notice that, while it is true in general that convergent sequences of points in a metric space satisfy the Cauchy condition, the converse does not always hold. The main reason for requiring that the metric space (X, d) be complete will be clear upon statement of the contraction mapping theorem, and specification of its role in the development of the theory. A familiar example of a complete metric space is given by $(\mathbb{R}^n, d)$, where d is the Euclidean metric on $\mathbb{R}^n$ defined by:

$$d(x, y) = \left\{ \sum_{i=1}^{n} (y_i - x_i)^2 \right\}^{1/2},$$

---

[1] Other recent references that we would like to suggest to the reader are Edgar (1990) and Peitgen et al. (1992). The former contains a lucid and rigorous exposition of topics in fractal geometry, including self-similarity. It also provides an excellent review of the necessary background material in metric topology and measure theory. The latter examines, at a simpler mathematical level, a wealth of topics in chaos theory and fractal geometry, and contains an abundance of material related to our discussion.

for any $x = (x_1, \ldots, x_n)^T$, $y = (y_1, \ldots, y_n)^T \in \mathbb{R}^n$.

Given a metric space (X, d), we can construct a new metric space $(\mathcal{H}(X), h)$, where $\mathcal{H}(X)$ is the collection of nonempty, compact subsets of X, and h is the Hausdorff metric on $\mathcal{H}(X)$ to be defined momentarily. Our interest in the collection $\mathcal{H}(X)$ is justified by the fact that it is extremely natural to identify images with closed and bounded subsets of the Euclidean plane. The Hausdorff metric, then, provides a means of measuring distances between such subsets. Its definition can be given as follows.

For A and B in $\mathcal{H}(X)$ and x in A, define first the distance from x to B as:

$$d(x, B) = \min_{y \in B} d(x, y).$$

Since $y \mapsto d(x, y)$ is a continuous function and B is compact and nonempty, the above minimum is actually attained. Next, define the distance from A to B as:

$$d(A, B) = \max_{x \in A} d(x, B).$$

For the same reason as above, this maximum is also attained. Note that, in general, $d(\cdot, \cdot)$ is not symmetric on $\mathcal{H}(X) \times \mathcal{H}(X)$, and therefore is not a metric. Finally, the Hausdorff distance between A and B is defined as:

$$h(A, B) = \max\left(d(A, B), d(B, A)\right).$$

A proof of the fact that $(\mathcal{H}(X), h)$ is indeed a metric space can be found in Barnsley (1988).

As an example, consider the metric space $(\mathbb{R}, d)$, where d is the Euclidean metric, so that, for any $x, y \in \mathbb{R}$, we have $d(x, y) = |y - x|$. Let $A = [0, 2]$ and $B = [1, 4]$. If $x \in [1, 2]$, we have $d(x, B) = 0$, while, if $x \in [0, 1)$, we have $d(x, B) = d(x, 1) = 1 - x > 0$. Thus, $d(A, B) = \max_{x \in [0,1)} d(x, 1) = \max_{x \in [0,1)}(1 - x) = 1 - 0 = 1$. On the other hand, if $y \in [1, 2]$, we have $d(y, A) = 0$. If $y \in (2, 4]$, we have $d(y, A) = d(y, 2) = y - 2 > 0$, so that $d(B, A) = \max_{y \in (2,4]} d(y, 2) = \max_{y \in (2,4]}(y - 2) = 4 - 2 = 2$. Hence, in this case, $d(A, B) = 1 \neq 2 = d(B, A)$, and $h(A, B) = \max(d(A, B), d(B, A)) = \max(1, 2) = 2$.

As shown in Barnsley (1988), the space $(\mathcal{H}(X), h)$ of nonempty, compact subsets of X has the very nice property of being complete, whenever (X, d) is. Furthermore, the limit A of a Cauchy sequence $\{A_n\}_{n=1}^{\infty}$ of points in $\mathcal{H}(X)$ can be characterized as:

$$A = \{x \in X | \exists \text{ a Cauchy sequence } \{x_n\}_{n=1}^{\infty} \text{ in } X \text{ s.t.}$$
$$x_n \in A_n \ \forall n, \text{ and } \lim_{n \to \infty} x_n = x\}.$$

It is also not difficult to prove that compactness of (X, d) implies compactness of $(\mathcal{H}(X), h)$.

2.2. Contractive and Affine Transformations

A very important class of transformations of one metric space into another, of which we shall make frequent use, consists of the collection of contractive transformations. Given two metric spaces (X, d_1) and (Y, d_2), a transformation $w : X \mapsto Y$ is said to be a *contraction* if and only if there exists a real number s, with $0 \le s \le 1$, such that:

$$d_2(w(x_1), w(x_2)) \le s d_1(x_1, x_2), \qquad \text{for any } x_1, x_2 \in X.$$

Any such number s is called a *contractivity factor* for w. If s is strictly less than one, w is said to be a *strict* contraction. Note that, when the two metric spaces coincide, w acts on pairs of points in X by bringing them closer together, their distance being reduced by a factor of at least s.

As an example, let $X = Y = \mathbb{R}$ and $d_1 = d_2 = d$, where d denotes the Euclidean metric on the real line. Consider a transformation $w : \mathbb{R} \mapsto \mathbb{R}$ of the form $w(x) = ax + b$, with a and b real constants. Then, for any $x_1, x_2 \in \mathbb{R}$, we have

$$d(w(x_1), w(x_2)) = |ax_2 + b - ax_1 - b| = |a||x_2 - x_1| = |a|d(x_1, x_2).$$

Hence, a necessary and sufficient condition for w to be a strict contraction (with contractivity factor $s = |a|$) is that $|a| < 1$.

Note that w is given by the composition of a linear transformation ax and a shift b. In general, given an $n \times n$ matrix A and a vector $b \in \mathbb{R}^n$, a transformation of $\mathbb{R}^n$ into itself of the form

$$w(x) = Ax + b, \qquad \text{for any } x \in \mathbb{R}^n,$$

is called *affine*.

In the sequel, we shall make extensive use of affine transformations, since they can be profitably employed as extremely manageable building blocks for constructing iterated function systems to be used for image generation. The main reason why affine transformations turn out to be so useful lies in their smoothness and regularity properties, and in the simplicity of affine arithmetic.

Consider, for example, the case of an affine transformation w of the Euclidean plane into itself, and suppose that $\det(A) \neq 0$, so that w is invertible. It is then easy to see that such a transformation is continuous, and maps lines into lines, and half-planes into half-planes. Furthermore, the image under w of any set in $\mathcal{H}(\mathbb{R}^2)$ will be a set in $\mathcal{H}(\mathbb{R}^2)$, whose shape will somewhat resemble the shape of the original set. This is so, since all affine transformations can do is rescale, translate, rotate, reflect, and shear the original set.

For instance, the image of the unit square under

$$ w\left((x_1, x_2)^T\right) = \left(\begin{array}{cc} 1/2 & 0 \\ 0 & 1/2 \end{array} \right) \left(\begin{array}{c} x_1 \\ x_2 \end{array} \right) $$

is the square of vertices

$$ (0,0)^T, \ (1/2,0)^T, \ (1/2,1/2)^T, \text{ and } (0,1/2)^T, $$

and the image of the unit square under

$$ w\left((x_1, x_2)^T\right) = \left(\begin{array}{cc} 1/2 & 0 \\ 0 & 1/4 \end{array} \right) \left(\begin{array}{c} x_1 \\ x_2 \end{array} \right) $$

is the rectangle of vertices

$$ (0,0)^T, \ (1/2,0)^T, \ (1/2,1/4)^T, \text{ and } (0,1/4)^T. $$

Another important fact, which proves useful in practical applications, is that an affine transformation of the plane into itself is uniquely determined by any three points and their images under the transformation, provided such points are not collinear. In fact, given any three pairs of preimage–image points, the six parameters of the transformation can be determined by solving a system of six linear equations.

Going back to the previous example, in order to determine an affine transformation

$$w\left((x_1, x_2)^T\right) = \begin{pmatrix} a & b \\ c & d \end{pmatrix} \begin{pmatrix} x_1 \\ x_2 \end{pmatrix} + \begin{pmatrix} e \\ f \end{pmatrix}$$

that maps the triangle of vertices $(0,0)^T$, $(1,0)^T$, and $(0,1)^T$ into the triangle of vertices $(0,0)^T$, $(1/2,0)^T$, and $(0,1/2)^T$, we can choose the six parameters a, b, c, d, e, and f in such a way that $w((0,0)^T) = (0,0)^T$, $w((1,0)^T) = (1/2,0)^T$, and $w((0,1)^T) = (0,1/2)^T$. The previous condition gives rise to the following system of linear equations:

$$\begin{aligned} a \cdot 0 + b \cdot 0 + e &= 0, \\ c \cdot 0 + d \cdot 0 + f &= 0, \\ a \cdot 1 + b \cdot 0 + e &= 1/2, \\ c \cdot 1 + d \cdot 0 + f &= 0, \\ a \cdot 0 + b \cdot 1 + e &= 0, \\ c \cdot 0 + d \cdot 1 + f &= 1/2, \end{aligned}$$

which admits the unique solution $a = d = 1/2$, $b = c = e = f = 0$. Hence, the required transformation is given, as expected, by:

$$w\left((x_1, x_2)^T\right) = \begin{pmatrix} 1/2 & 0 \\ 0 & 1/2 \end{pmatrix} \begin{pmatrix} x_1 \\ x_2 \end{pmatrix}.$$

Suppose now that $w(x) = Ax + b$ is an affine transformation of $\mathbb{R}^n$, endowed with the Euclidean metric d, into $\mathbb{R}^n$, endowed with the same metric. Then, for any two points x_1, and x_2 in $\mathbb{R}^n$, we have:

$$\begin{aligned} d(w(x_1), w(x_2)) &= |Ax_2 + b - Ax_1 - b| = |A(x_2 - x_1)| \\ &\le \|A\| \, |x_2 - x_1| = \|A\| \, d(x_1, x_2), \end{aligned}$$

where the vector norm $|\cdot|$ is the one induced by the Euclidean metric, and

$$\|A\| = \max\left\{ \frac{|Ax|}{|x|} \,\middle|\, x \ne (0,0)^T \right\}$$

defines the norm of A, and coincides with $\lambda_{\max}$, the square root of the largest eigenvalue of $A^T A$. The maximum in the definition is achieved at the corresponding eigenvector of $A^T A$ (see Strang (1980)). It follows that a necessary and sufficient condition for w to be a strict contraction (with contractivity factor $s = \lambda_{\max}$) is that the square root of the largest eigenvalue of $A^T A$ be strictly less than one.

For the affine transformation of $\mathbb{R}^2$ into itself

$$w\left((x_1, x_2)^T\right) = \begin{pmatrix} 1/2 & 0 \\ 0 & 1/2 \end{pmatrix} \begin{pmatrix} x_1 \\ x_2 \end{pmatrix},$$

we have

$$A^T A = A^2 = \begin{pmatrix} 1/4 & 0 \\ 0 & 1/4 \end{pmatrix},$$

so that the square root of the largest eigenvalue of $A^T A$ is given by $\sqrt{1/4} = 1/2$. Hence, w is a contraction of the Euclidean plane into itself, with contractivity factor $s = 1/2$.

The following theorem, known as the *contraction mapping theorem*, and whose proof can be found in Barnsley (1988), states an important property of contractive transformations of a complete metric space into itself.

Theorem 2.1. *Let $w : X \mapsto X$ be a strict contraction on a complete metric space (X, d). Then, there exists a unique point $x_f \in X$ such that $w(x_f) = x_f$. Furthermore, for any $x \in X$, we have $\lim_{n \to \infty} w^{\circ n}(x) = x_f$, where $w^{\circ n}$ denotes the n-fold composition of w with itself.*

The contraction mapping theorem is remarkable for two reasons. On the one hand, it guarantees the existence of a unique *fixed point* x_f for the contraction w, and, on the other, it suggests how such a fixed point can be computed iteratively. The contraction

$$w\left((x_1, x_2)^T\right) = \begin{pmatrix} 1/2 & 0 \\ 0 & 1/2 \end{pmatrix} \begin{pmatrix} x_1 \\ x_2 \end{pmatrix}$$

of the Euclidean plane into itself, for instance, has the origin as its unique fixed point. In this case, since the fixed point can be easily computed analytically, the second part of the theorem does not seem to bear much relevance. However, we shall soon encounter situations in which the iterative procedure constitutes a viable alternative to a difficult and sometimes impossible analytic determination of the fixed point.

2.3. Iterated Function Systems

In this section, we shall introduce the notion of deterministic iterated function systems, and show how they can be employed to encode digitized images. The fundamental ideas were first exposed and developed by Hutchinson (1981) in an extremely interesting and stimulating paper. For proofs of the results that we are going to outline, we also refer the reader to the somewhat simpler and more detailed exposition provided in Barnsley (1988).

Basically, given a complete metric space (X, d), we shall consider the associated space $(\mathcal{H}(X), h)$ of nonempty, compact subsets of X, endowed with the Hausdorff metric, and define a contractive transformation W of $\mathcal{H}(X)$ into itself. By the contraction mapping theorem, W will have a unique fixed point in $\mathcal{H}(X)$. In particular, for the case $X = \mathbb{R}^2$, this construction will enable us to identify certain images (nonempty, compact subsets of the Euclidean plane) with fixed points of contractive transformations of $\mathcal{H}(\mathbb{R}^2)$ into itself. The single most important by-product of this identification, which has far-reaching consequences in the area of image encoding, lies in the possibility of obtaining considerable reduction in the amount of data needed to describe digitized images.

We start by considering a metric space (X, d) and a finite set of strictly contractive transformations $w_n : X \mapsto X$, $1 \leq n \leq N$, with respective contractivity factors s_n. We then proceed to define a transformation $W : \mathcal{H}(X) \mapsto \mathcal{H}(X)$, where $\mathcal{H}(X)$ is the collection of nonempty, compact subsets of X, by:

$$W(B) = \bigcup_{n=1}^{N} w_n(B), \qquad \text{for any } B \in \mathcal{H}(X). \qquad (2.1)$$

It is easily shown (see Barnsley (1988)) that W is a strict contraction, with contractivity factor $s = \max_{1 \leq n \leq N} s_n$. It follows from the contraction mapping theorem that, if (X, d) is complete, W has a unique fixed point A in $\mathcal{H}(X)$, satisfying the remarkable self-covering condition

$$A = W(A) = \bigcup_{n=1}^{N} w_n(A). \tag{2.2}$$

We now give the following definition.

Definition 2.2. *A hyperbolic iterated function system (IFS) $\{X; w_1, \ldots, w_n\}$ consists of a complete metric space (X, d) and a finite set of strictly contractive transformations $w_n : X \mapsto X$ with contractivity factors s_n, for $n = 1, \ldots, N$. The maximum s among $s_1, \ldots, s_N$ is called a* contractivity factor *for the IFS. The unique fixed point in $\mathcal{H}(X)$ of the transformation W defined by relation (2.1) is called the* attractor *of the IFS.* [2]

As a first, simple example, consider the IFS $\{\mathbb{R}; w_1(x) = (1/2)x, w_2(x) = (1/2)x + 1/2\}$ on the complete metric space given by the real line endowed with the Euclidean metric. Both w_1 and w_2 are contractions, each with contractivity factor $1/2$, so that $1/2$ is also a contractivity factor for the IFS. Let us guess that the attractor of the IFS is a closed interval $A = [a, b]$. Then we have $w_1(A) = [(1/2)a, (1/2)b]$ and $w_2(A) = [(1/2)a + 1/2, (1/2)b + 1/2]$ so that, in order for A to be the attractor, the following condition must be satisfied:

$$[a, b] = \left[\frac{1}{2}a, \frac{1}{2}b\right] \bigcup \left[\frac{1}{2}a + \frac{1}{2}, \frac{1}{2}b + \frac{1}{2}\right].$$

It is therefore necessary that $a = (1/2)a$ and $b = (1/2)b + 1/2$, which implies $a = 0$ and $b = 1$. For these values of a and b, we obtain $(1/2)b = 1/2 = (1/2)a + 1/2$, so that

$$[0, 1] = \left[0, \frac{1}{2}\right] \bigcup \left[\frac{1}{2}, 1\right] = w_1([0, 1]) \bigcup w_2([0, 1]) = W([0, 1]).$$

Thus, $A = [0, 1]$, being the unique fixed point of W, is the attractor of the IFS.

---

[2]While Hutchinson (1981) is an earlier reference on the topic, the term IFS was first introduced, to our knowledge, in Barnsley and Demko (1985). In our definition, we have followed closely Barnsley (1988), whose notation we have also adopted. We like the fact that it indicates both the transformations comprising the IFS and the metric space on which they act. Some authors (Edgar (1990), for example) refer to a hyperbolic IFS as a contractive IFS.

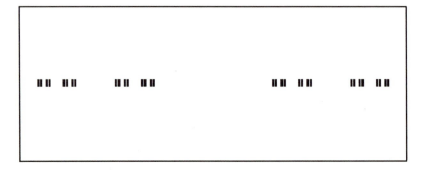

Figure 2.1. Cantor set.

Here, determination of the attractor has been decisively aided by the initial guess about its being a closed interval. Obviously, this cannot be considered a reasonable way to proceed in the general case, when there are several transformations and nothing can be easily inferred from them about the shape of the attractor. A more sensible way to approach the problem of determining the attractor of a given IFS is based on the second part of the contraction mapping theorem. Basically, one would start with any nonempty compact set and follow its evolution under iterations by W, the limiting set being the attractor of the IFS.

For instance, if we take the initial set to be $A_0 = [0, 2]$ and apply W, we obtain:

$$\begin{aligned} A_1 &= W(A_0) = w_1(A_0) \bigcup w_2(A_0) \\ &= [0, 1] \bigcup \left[\frac{1}{2}, 1 + \frac{1}{2}\right] = \left[0, 1 + \frac{1}{2}\right]. \end{aligned}$$

If we keep iterating W, we have, at stage n, $A_n = [0, 1 + (1/2^n)]$, and it is easy to verify that the sequence $\{[0, 1 + (1/2^n)]\}_{n=0}^{\infty}$ admits the interval $[0, 1]$ as its unique limit within the space of nonempty, compact subsets of the real line, endowed with the Hausdorff metric induced by the Euclidean metric on $\mathbb{R}$.

Another interesting example of an IFS on the real line is given by $\{\mathbb{R}; w_1(x) = (1/3)x, \ w_2(x) = (1/3)x + 2/3\}$, whose attractor is given by the Cantor set of Figure 2.1. This set can be obtained iteratively, starting with the unit interval, and successively removing the middle open third of each interval that is left at the previous stage of the construction. More precisely, letting $E_0 = [0, 1]$, we construct E_1 by removing the open segment $(1/3, 2/3)$, so that $E_1 = [0, 1/3] \cup [2/3, 1]$. Next, we construct E_2 by removing the two open segments $(1/9, 2/9)$ and $(7/9, 8/9)$ from E_1, thus obtaining $E_2 = [0, 1/9] \cup [2/9, 1/3] \cup [2/3, 7/9] \cup [8/9, 1]$. Proceeding in this fashion, we construct a nested sequence, $E_0 \supset E_1 \supset E_2 \supset \ldots$, of compact subsets of the unit interval. The Cantor set is then defined as $E = \bigcap_{k=0}^{\infty} E_k$. It can be shown (see Rudin (1976)) that E is nonempty, compact and perfect, thus containing an uncountable number of points. In order to show that the attractor of the IFS in our example actually coincides with the Cantor set, we shall make use of the following result, which we state without proof. [3]

Lemma 2.3. *Let (X, d) be a complete metric space, and let $\{X; w_1, \ldots, w_N\}$ be a hyperbolic IFS with attractor A. Given any set $A_0 \in \mathcal{H}(X)$, define:*

$$A_k = W^{\circ k}(A_0), \qquad \text{for } k = 0, 1, 2, \ldots \ .$$

The following results hold.

(i) If $A_0 \subset A_1$, then $A_0 \subset A_1 \subset A_2 \subset \ldots \subset A$, and $A = \overline{\bigcup_{k=0}^{\infty} A_k}$, where the overlying bar denotes set closure.

(ii) If $A_0 \supset A_1$, then $A_0 \supset A_1 \supset A_2 \supset \ldots \supset A$, and $A = \bigcap_{k=0}^{\infty} A_k$.

Returning to our example, if we take $A_0 = E_0 = [0, 1]$, we have $w_1(A_0) = [0, 1/3]$ and $w_2(A_0) = [2/3, 1]$, so that $A_1 = W(A_0) = E_1$. At the next step, we have $w_1(A_1) = [0, 1/9] \cup [2/9, 1/3]$ and $w_2(A_1) = [2/3, 7/9] \cup [8/9, 1]$, so that $A_2 = W^{\circ 2}(A_0) = E_2$. In general, for any nonnegative integer k, we shall have $A_k = W^{\circ k}(A_0) = E_k$. Since $A_0 \supset A_1$, we can apply part (ii) of Lemma 2.3, and conclude that the Cantor set is indeed the attractor of the given IFS.

A nice two-dimensional analogue of the Cantor set is given by the Sierpinski triangle of Figure 2.2, which can also be obtained as the attractor of an IFS. The Sierpinski triangle is similar to the Cantor

---

[3] We learned this lemma, in its stated form, from Steven Shreve (see Acknowledgments). Its proof follows easily from results in Hutchinson (1981) and Barnsley (1988).

Figure 2.2. Sierpinski triangle.

set in that, as one can see from the illustration, it can be constructed by successively removing open triangular subsets from the closed triangle of vertices $(0,0)^T$, $(1,0)^T$, and $(1/2,1)^T$. One possible IFS on the Euclidean plane that admits this Sierpinski triangle as its attractor is given by $\{\mathbb{R}^2; w_1, w_2, w_3\}$, where

$$w_1\left((x_1, x_2)^T\right) = \begin{pmatrix} 0.5 & 0 \\ 0 & 0.5 \end{pmatrix}\begin{pmatrix} x_1 \\ x_2 \end{pmatrix},$$

$$w_2\left((x_1, x_2)^T\right) = \begin{pmatrix} 0.5 & 0 \\ 0 & 0.5 \end{pmatrix}\begin{pmatrix} x_1 \\ x_2 \end{pmatrix} + \begin{pmatrix} 0.5 \\ 0 \end{pmatrix},$$

$$w_3\left((x_1, x_2)^T\right) = \begin{pmatrix} 0.5 & 0 \\ 0 & 0.5 \end{pmatrix}\begin{pmatrix} x_1 \\ x_2 \end{pmatrix} + \begin{pmatrix} 0.25 \\ 0.5 \end{pmatrix}.$$

Also in this case, part (ii) of Lemma 2.3 can be used, starting with the

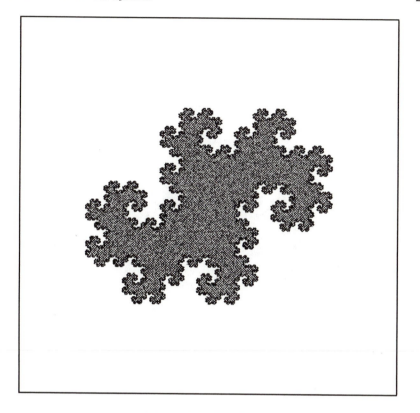

Figure 2.3. Twin dragon.

closed triangle of vertices $(0,0)^T$, $(1,0)^T$, and $(1/2,1)^T$, to convince ourselves that the Sierpinski triangle is the attractor of the given IFS.

Other examples of images that can be obtained as attractors of IFSs on the Euclidean plane are provided by Figures 2.3 – 2.9. Figure 2.3 represents an object known in the literature as the twin dragon. It constitutes the attractor of the IFS $\{\mathbb{R}^2; w_1, w_2\}$, where

$$w_1\left((x_1,x_2)^T\right) = \begin{pmatrix} 0.5 & 0.5 \\ -0.5 & 0.5 \end{pmatrix}\begin{pmatrix} x_1 \\ x_2 \end{pmatrix} + \begin{pmatrix} 0.125 \\ 0.625 \end{pmatrix},$$

$$w_2\left((x_1,x_2)^T\right) = \begin{pmatrix} 0.5 & 0.5 \\ -0.5 & 0.5 \end{pmatrix}\begin{pmatrix} x_1 \\ x_2 \end{pmatrix} + \begin{pmatrix} -0.125 \\ 0.375 \end{pmatrix}.$$

The maple leaf of Figure 2.4 can be obtained as the attractor of

Figure 2.4. Maple leaf.

the IFS $\{\mathbb{R}^2; w_1, \ldots, w_4\}$, where

$$w_1\left((x_1, x_2)^T\right) = \begin{pmatrix} 0.8 & 0 \\ 0 & 0.8 \end{pmatrix}\begin{pmatrix} x_1 \\ x_2 \end{pmatrix} + \begin{pmatrix} 0.1 \\ 0.04 \end{pmatrix},$$

$$w_2\left((x_1, x_2)^T\right) = \begin{pmatrix} 0.5 & 0 \\ 0 & 0.5 \end{pmatrix}\begin{pmatrix} x_1 \\ x_2 \end{pmatrix} + \begin{pmatrix} 0.25 \\ 0.4 \end{pmatrix},$$

$$w_3\left((x_1, x_2)^T\right) = \begin{pmatrix} 0.355 & -0.355 \\ 0.355 & 0.355 \end{pmatrix}\begin{pmatrix} x_1 \\ x_2 \end{pmatrix} + \begin{pmatrix} 0.266 \\ 0.078 \end{pmatrix},$$

$$w_4\left((x_1, x_2)^T\right) = \begin{pmatrix} 0.355 & 0.355 \\ -0.355 & 0.355 \end{pmatrix}\begin{pmatrix} x_1 \\ x_2 \end{pmatrix} + \begin{pmatrix} 0.378 \\ 0.434 \end{pmatrix}.$$

Figure 2.5 represents a black spleenwort fern, and it can be gen-

Figure 2.5. Black spleenwort fern.

erated as the attractor of the IFS $\{\mathbb{R}^2; w_1, \ldots, w_4\}$, where

$$w_1\left((x_1, x_2)^T\right) = \begin{pmatrix} 0.856 & 0.0414 \\ -0.0205 & 0.858 \end{pmatrix}\begin{pmatrix} x_1 \\ x_2 \end{pmatrix} + \begin{pmatrix} 0.07 \\ 0.147 \end{pmatrix},$$

$$w_2\left((x_1, x_2)^T\right) = \begin{pmatrix} 0.244 & -0.385 \\ 0.176 & 0.224 \end{pmatrix}\begin{pmatrix} x_1 \\ x_2 \end{pmatrix} + \begin{pmatrix} 0.393 \\ 0.102 \end{pmatrix},$$

$$w_3\left((x_1, x_2)^T\right) = \begin{pmatrix} -0.144 & 0.39 \\ 0.181 & 0.259 \end{pmatrix}\begin{pmatrix} x_1 \\ x_2 \end{pmatrix} + \begin{pmatrix} 0.527 \\ -0.014 \end{pmatrix},$$

$$w_4\left((x_1, x_2)^T\right) = \begin{pmatrix} 0 & 0 \\ 0.031 & 0.216 \end{pmatrix}\begin{pmatrix} x_1 \\ x_2 \end{pmatrix} + \begin{pmatrix} 0.486 \\ 0.05 \end{pmatrix}. \quad [4]$$

---

[4]The IFS encodings of the maple leaf and of the fern are both due to Barnsley.

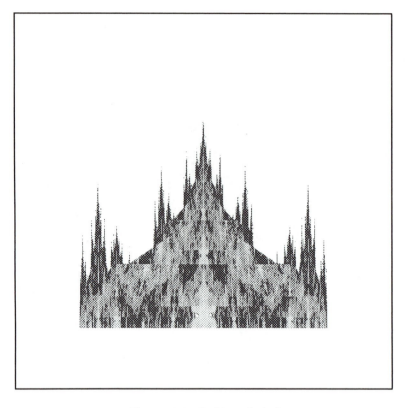

Figure 2.6. Gothic cathedral.

An appropriate window for displaying the attractors of all these IFSs is given by the unit square $[0, 1] \times [0, 1]$. The Gothic cathedral of Figure 2.6 and Color Plate 1 (vaguely resembling the Duomo of Milan), the galaxy of Figure 2.7 and Color Plate 2, the pine tree of Figure 2.8 and Color Plate 3, and the spiral staircase of Figure 2.9 are also attractors of IFSs on the Euclidean plane, all consisting of at most 16 affine transformations.

As the examples show, the shapes of images that can be obtained as attractors of IFSs are quite varied and complicated. What is most fascinating is that all the information needed to describe such images is contained in the parameters of the affine transformations. The Sierpinski triangle, for instance, is uniquely determined by 18 numbers, while 12 numbers are all that is needed to pin down the twin dragon. This is extremely surprising, if one considers that the

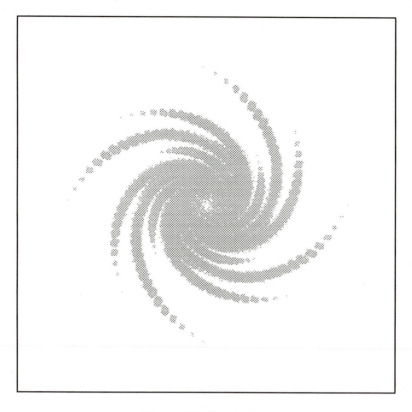

Figure 2.7. Spiral galaxy.

elaborateness of the boundary of the dragon, or of any of the other images for that matter, makes it almost impossible to describe them in words.

The reduction in the amount of data required to code these images is what makes IFS theory potentially valuable in practical applications. In order to describe an image that can be regarded as the attractor of an IFS, one only needs to keep track of the transformations in the IFS. For such an image, the attainable level of data compression is usually spectacular and, in general, far superior to what would be achieved by more traditional compression techniques.

Of course, in order for a compression method based on IFS theory to be easily and profitably employed, two fundamental issues ought to be addressed. First, one needs a way, given an arbitrary image, of finding an IFS whose attractor coincides with, or closely approxi-

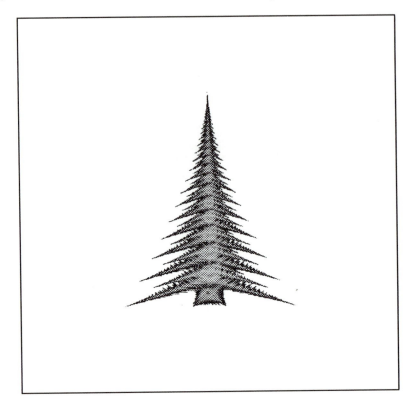

Figure 2.8. Pine tree.

mates, that image. This problem, known as the encoding or inverse problem, is inherently difficult. Several computationally intensive solutions have been proposed in the literature, and we shall discuss a few of them in the sequel. In addition, we shall briefly outline, at the end of Chapter 4, some ideas and preliminary work related to a new approach that we are in the process of developing.

The second issue that has to be addressed is related to the availability of an efficient way of decoding images. In other words, one needs algorithms to perform speedy and accurate generation of attractors of IFSs. As previously noted, one such algorithm could be based on the convergence result stated in the second part of the contraction mapping theorem. Such an algorithm, however, turns out to be more difficult to implement and less efficient than a probabilistic algorithm, known as the *random iteration algorithm,* which will be

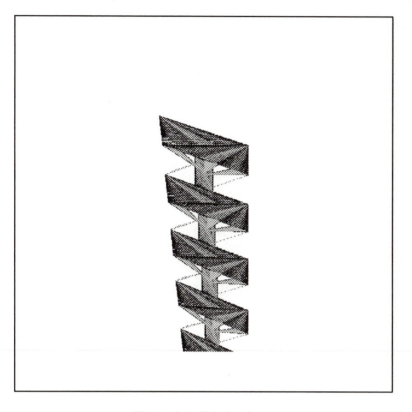

Figure 2.9. Spiral staircase.

introduced in the next chapter.

As a matter of fact, every attractor presented in this book has been generated by means of the random iteration algorithm, or one of its variations. The probabilistic algorithm is remarkable as far as ease of implementation and speed of execution are concerned, and offers the additional advantage of providing a simple and natural way of adding color to the images that are generated, as we shall see in the sequel. [5]

---

[5]The properties and relative merits of various methods for generating the attractor of an IFS are carefully examined in Hepting et al. (1991). In addition, the article presents methods for the interesting, related problem of rendering the complement of the attractor.

2.4.   Iterated Function Systems with Condensation

Iterated function systems admit a simple and useful generalization, if one allows one of the constituent functions to be a degenerate transformation of the space of nonempty, compact subsets of the metric space under consideration into itself. The generalization, introduced in Barnsley and Demko (1985), is useful, since it broadens the class of images that can be identified with attractors of dynamical systems, and offers the possibility of obtaining a higher degree of control over the encoding procedure.

Formally, given a complete metric space (X, d), we choose a fixed, nonempty compact set $C \in \mathcal{H}(X)$, and define $w_0 : \mathcal{H}(X) \mapsto \mathcal{H}(X)$ by $w_0(B) = C$, for any $B \in \mathcal{H}(X)$. In other words, w_0 maps every set belonging to $\mathcal{H}(X)$ into the same set C. We then give the following definition.

Definition 2.4. *The collection* $\{X; w_0, \ldots, w_N\}$, *where* $w_1, \ldots, w_N$ *are strict contractions on* X, *and* w_0 *is defined as above, is called a* hyperbolic iterated function system with condensation. *The set* C *that appears in the definition of* w_0 *is called the* condensation set. [6]

It is proved in Barnsley (1988) that $W : \mathcal{H}(X) \mapsto \mathcal{H}(X)$, defined by

$$W(B) = \bigcup_{n=0}^{N} w_n(B), \qquad \text{for any } B \in \mathcal{H}(X),$$

is a contraction on $\mathcal{H}(X)$, with respect to the Hausdorff metric h induced by the metric d on X. Hence, since $(\mathcal{H}(X), h)$ is complete, the contraction mapping theorem implies the existence of a unique set $A \in \mathcal{H}(X)$ such that $W(A) = A$. Such a set A is called the *attractor* of the IFS with condensation, and it satisfies a property which is a direct generalization of the self-covering property for regular IFSs.

---

[6]Here again, as in the case of Definition 2.2, we have adopted the notation and terminology used in Barnsley (1988).

Figure 2.10. Mixing of Sierpinski triangle and maple leaf.

More precisely, there holds:

$$A = \bigcup_{n=0}^{N} w_n(A) = C \bigcup \left(\bigcup_{n=1}^{N} w_n(A) \right), \qquad (2.3)$$

so that the attractor is given by the union of N images of itself under $w_1, \ldots, w_N$, and of the condensation set C.

Figures 2.10 and 2.11 (see also Color Plates 4 and 8) provide two examples of attractors of IFSs with condensation. In Figure 2.10, the condensation set is given by the maple leaf within the middle portion of the Sierpinski triangle, so that w_0 maps every nonempty, compact subset of the plane into the leaf. The three additional transformations that form the IFS are the same transformations that generate the Sierpinski triangle of Figure 2.2. Note that, now, the

Figure 2.11. Row of pine trees.

resulting image is no longer self-covering in the way that attractors of regular IFSs are, since the condensation set itself is needed to obtain a complete covering.

Figure 2.11 depicts a row of pine trees of different heights. The condensation set is given by the tallest tree on the left, and the IFS is completed by the additional affine transformation

$$w_1\left((x_1, x_2)^T\right) = \begin{pmatrix} 0.6 & 0 \\ 0 & 0.6 \end{pmatrix} \begin{pmatrix} x_1 \\ x_2 \end{pmatrix} + \begin{pmatrix} 0.4 \\ 0 \end{pmatrix},$$

which linearly attracts points in the plane towards its fixed point $(1, 0)^T$.

When the condensation set, as it is the case in these examples, is itself the attractor of an IFS, a modification of the random iteration algorithm mentioned in the previous section provides an effective and quick way of generating the attractor of an IFS with condensation. We shall explain the details of the algorithm in the following chapter.

2.5.   Self-Covering and Encoding

The possibility of applying IFS techniques as an aid to image data compression is, at the moment, limited. One basic problem has to be fully addressed and satisfactorily solved before such methods can be widely and routinely employed. As we shall see, once the transformations of an IFS have been provided, the generation (or *decoding*) of the corresponding image through the random iteration algorithm constitutes a quick and easy task. However, in most practical situations, the image itself is all that is given, so that the key issue consists in being able to determine transformations that will generate such an image. It should be noticed that the problem, as just stated, is not well-posed. In fact, there is no guarantee that, for an arbitrarily given image, there exists an IFS that will generate it, nor that, when such an IFS exists, it is unique. However, there are several possible ways of slightly modifying the original problem and making it precise. Any of these modifications of the original problem is usually referred to, in the literature, as an *inverse* or *encoding* problem.

In this section, we review some of the ideas presented in Berger (1989a). We shall deal exclusively with black-and-white images, since color images, being identifiable with probability distributions, best fit into the probabilistic framework that will be addressed in the following chapter. One simple and obvious way of posing a sensible inverse problem is to start by considering a black-and-white image generated by an IFS consisting of N contractive transformations of $\mathbb{R}^2$ into itself. Then, knowing that the image can actually be generated by contractive transformations, the ultimate goal becomes that of being able to determine such transformations. All the available information is represented, in this case, by the shape of the image and by the knowledge of the number of transformations employed to generate it.

Different variations of this basic inverse problem can also be considered. For instance, a more difficult problem arises when the number of transformations employed is also unknown, in which case one would like to be able to determine the minimum number of transformations necessary to generate the given image. Easier versions of the problem are those in which more partial information is provided. For example, some of the transformations may be known.

All inverse problems described so far share the common characteristic of dealing with images that are known to have been generated through IFSs consisting of contractive transformations. When the image to be reproduced is completely arbitrary, a new class of inverse problems arises, the goal becoming that of finding images that can be generated by means of IFSs and that constitute good approximations to the original image.

The approximation process requires the specification of some criteria for measuring the distance between various images. Since black-and-white images can be regarded as subsets of $\mathbb{R}^2$, reasonable measures of the distance between two images can be obtained by considering their Hausdorff distance, or the area of their symmetric difference.

An important observation has to be made regarding the degree of accuracy that can be achieved when trying to reconstruct a given image. If one does not set any limit on the number of transformations to be employed, then, in view of the self-covering property (2.2), any prespecified degree of approximation can be obtained. This can simply be done by covering the image with a very large number of little copies of itself under contractive transformations. By doing so, however, no data compression would be achieved. A more sensible way of stating the problem, instead, is to specify a desired degree of accuracy (measured, for instance, in terms of any of the suggested distances) and to try to determine the least number of contractive transformations necessary to reproduce the given image within the required approximation. Alternatively, one could try to evaluate the best possible degree of approximation attainable when employing a prespecified number of transformations.

The techniques that have so far been devised to tackle the inverse problem can be employed both in the case when the given image can be exactly reproduced by means of an IFS and in the case when it can only be approximated. In order to simplify the discussion, we shall restrict our attention, throughout the remainder of the section, to IFSs consisting only of affine transformations.

A conceptually simple and interesting approach to the solution of the inverse problem is based on the self-covering property (2.2), and it can be easily illustrated by means of an example. Consider the Sierpinski triangle of Figure 2.2, which has been generated from the three mappings listed on page 20. In order to recover the transformations from the image, we can try to determine how the triangle can be covered with affine copies of itself. In this particular case, it is easy to see that the image is the union of three affine copies of itself. More precisely, as shown in Figure 2.12, the original triangle with vertices (A, B, C) is the just-touching union of the three smaller triangles with vertices (A, D, F), (D, B, E), and (F, E, C), respectively. [7] As we have observed in Section 2.2, an affine transformation of $\mathbb{R}^2$ into itself is uniquely determined by any three points and their images under the transformation, provided such points are not collinear. Hence, w_1 may be characterized as the unique affine transformation that maps A, B, and C into A, D, and F, respectively, and the actual calculation of the parameters of w_1 only requires the solution of a simple system of linear equations. Likewise, the parameters of w_2 and w_3 can be easily determined.

In summary, this procedure amounts to finding a covering of the original image with affine copies of itself, each copy corresponding to a transformation of the IFS that has to be recovered. An important constraint to be made is that all the obtained transformations must be strictly contractive. This rules out, for instance, the solution where all the mappings coincide with the identity, although the self-covering condition is trivially satisfied in that case.

The extremely simple nature of the example used to illustrate the method is not revealing of the difficulties that can be encountered in practical applications. In particular, since the elements of the covering need not be disjoint, it may be difficult, in general, to determine what they are. The images of Figures 2.4 and 2.5 offer less trivial examples of the self-covering property. As shown in Figure 2.13, the maple leaf can be covered with four *overlapping* affine copies of itself. Figure 2.14 shows how the fern can also be covered with four

---

[7]Here, we have used the term "just-touching" informally. The reader can find its rigorous definition in terms of addresses in code space in Barnsley (1988).

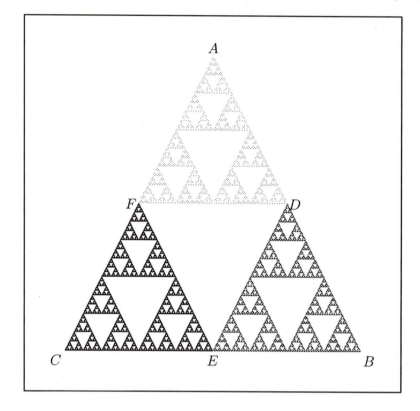

Figure 2.12. Self-covering of the Sierpinski triangle.

copies of itself. Notice, however, that one of these copies (the stem of the fern) does not resemble the shape of the whole image and corresponds to a degenerate map.

When the given image cannot be exactly reproduced, the inverse problem can be viewed as a minimization problem in a large number of variables. The theoretical grounds on which such an optimization problem rests are provided by the following result, known as the *collage theorem,* whose proof can be found in Barnsley (1988).

Theorem 2.5. *Let (X, d) be a complete metric space, and let $T \in \mathcal{H}(X)$ and $\epsilon > 0$ be given. Suppose that a hyperbolic IFS with attractor A and contractivity factor s is such that:*

$$h(T, W(T)) \leq \epsilon, \qquad (2.4)$$

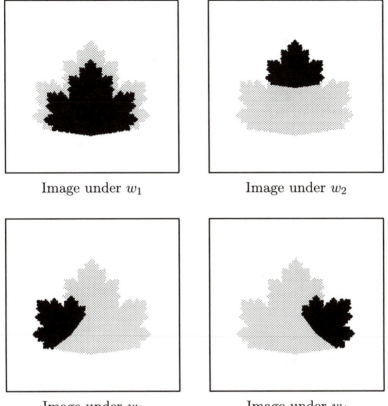

Image under w_1 Image under w_2

Image under w_3 Image under w_4

Figure 2.13. Self-covering of the maple leaf.

where h denotes the Hausdorff metric on $\mathcal{H}(X)$. Then:

$$h(T, A) \leq \frac{\epsilon}{1 - s}. \qquad (2.5)$$

Suppose, then, that a target image T is given, and imagine that we want to find an IFS consisting of N affine transformations, whose attractor approximates T. (Here, for simplicity, we assume that the number of transformations has been prespecified.) The collage theorem tells us that, if we can find a hyperbolic IFS with contractivity factor s such that the union of the images of T under the transformations in the IFS covers T itself to within a distance of ϵ, then the

Image under w_1

Image under w_2

Image under w_3

Image under w_4

Figure 2.14. Self-covering of the black spleenwort fern

distance of the attractor of the IFS from the target image will be bounded above by $\epsilon/(1 - s)$. In other words, if we can construct an IFS that provides a good approximate self-covering of the target image, the attractor of such an IFS will also be a good approximation to the target.

Notice that matters are slightly complicated by the presence of the factor $(1 - s)$ in the denominator of the right-hand side of inequality (2.5). In fact, it is clear from inequalities (2.4) and (2.5) that, in order for the collage theorem to guarantee that the attractor of the chosen IFS is a good approximation to the target image, not only does the left-hand side of inequality (2.4) have to be small, but also the IFS must have a contractivity factor which is not too close to one.

Hence, a practical way of solving the problem is to try to minimize the left-hand side of inequality (2.4), regarded as a function of the parameters of the affine transformations in the IFS, subject to the condition that the IFS has a contractivity factor which is bounded away from one. Currently, this rather difficult minimization problem is being solved by stochastic relaxation methods, based on simulated annealing. Some of the reasons why the method is difficult to implement are the large number of variables usually involved, the presence of the aforementioned constraint, the existence of several local minima for the objective function, and the fact that each function evaluation requires the time-consuming calculation of the Hausdorff distance between two subsets of $\mathbb{R}^2$. [8]

---

[8]Recently, significant advances have been made in the direction of finding an automatic method for determining an IFS encoding of a given image (see Jacquin (1992) and Barnsley and Hurd (1993)). The method is based on the self-covering property for local IFSs. These are IFSs consisting of local contractive transformations, i.e., transformations that are contractive on a subset of the original metric space. The efficacy of the method stems from the fact that an encoding for the whole image can be computed by working separately on easier encoding subproblems for small subsets of the image.

3

Probabilistic Theory of Iterated Function Systems

The present chapter will be devoted to the development of IFS theory from a probabilistic point of view, which departs significantly from the approach taken in the previous chapter. The focus will be kept, once again, on techniques for representing digitized images. However, while we have so far identified images with fixed points of certain contractive transformations of the space of nonempty, compact subsets of the Euclidean plane into itself, we shall now describe ways in which the same images can be identified with probability distributions.

There are two important advantages connected to this identification. First, introduction of a correspondence between frequencies and color intensities provides a natural means of dealing with color images. This represents a significant step forward compared to the techniques outlined in the previous chapter, which only allowed treatment of black-and-white images. Furthermore, the probabilistic setting provides the theoretical background for an iterative algorithm that can be employed to rapidly generate both black-and-white and color images. Such an algorithm turns out to be much more efficient than a possible deterministic algorithm based on the convergence result provided by the contraction mapping theorem.

For the purpose of illustration, let us restrict our attention to the case of the Euclidean plane. The basic idea consists in associating a positive probability to every transformation in a given IFS on $\mathbb{R}^2$, thereby constructing what is known as an IFS with probabilities, and using this new object to define the transition probability structure of a Markov chain on the plane. Under certain conditions on the transformations and the associated probabilities, the resulting Markov chain possesses a unique stationary distribution that can be identified with a color image, while the support of such a distribution can be regarded as a black-and-white image. In particular, when the conditions stated in the previous chapter are satisfied, namely, when the IFS is hyperbolic, the support of the stationary distribution coincides with the attractor of the IFS, as defined in Definition 2.2.

The most important feature of the resulting Markov chain is that the empirical distributions of almost every trajectory converge weakly to the unique stationary distribution of the chain. This suggests a way of generating the associated image by running a trajectory of the Markov chain and plotting the points as they are generated. The aforementioned property guarantees that the resulting image will always be the same, irrespective of the particular trajectory that is being generated.

The probabilistic setting also provides a way of weakening the assumptions on the transformations that comprise the IFS, while still preserving the essential property that the given IFS has a unique associated image. Since the contraction mapping theorem is no longer employed in proving the results, one need not require that all transformations in the IFS be contractive; an appropriate notion of average contractivity will suffice. However, when the IFS is hyperbolic, the contraction mapping theorem can still be employed in proving the existence of a unique invariant probability for the Markov chain, once a suitable notion of distance on the space of probabilities has been defined.

We shall first describe the basic generation algorithm and introduce a stochastic model, based on affine transformations, that provides its theoretical background, discussing briefly a set of conditions that can be imposed on the transformations and the probabilities in order for the algorithm to work. We shall then describe the self-covering property for the resulting images in the present probabilistic setting, and make some remarks on the encoding problem for full-color images. An extension of the basic algorithm, based on a mixing technique, and its connection to IFSs with condensation will also be considered. We shall then indicate how a fixed-point argument can still be used in order to prove the main results. Finally, we shall conclude the chapter by describing some theoretical results we were able to obtain by applying the invariance condition to the stationary distributions associated with certain IFSs with probabilities, and derive a novel characterization of all Beta distributions with integer parameters in terms of IFSs with probabilities.

3.1. Generation Algorithm

The algorithm for image generation described in this section is due to Barnsley (1988), and is based, as we have already mentioned, on IFS theory. Since, for the practical purpose of generating digitized images, it is often convenient to consider IFSs based only on affine transformations of the plane into itself, the theoretical discussion here will be restricted to such a framework. A more general and thorough development of this theory will be presented in the following sections.

Consider N affine transformations of $\mathbb{R}^2$ into itself of the form:

$$w_i(x) = A_i x + b_i, \qquad 1 \le i \le N,$$

and associate to each of them a positive probability p_i, with $\sum_{i=1}^N p_i = 1$. In analogy with Definition 2.2, $\{\mathbb{R}^2; w_1, \ldots, w_N; p_1, \ldots, p_N\}$ is called an IFS with probabilities. The recursion

$$X_{n+1} = w_{\sigma_n}(X_n), \qquad n = 0, 1, 2, \ldots, \tag{3.1}$$

where σ_n are independent integer random variables such that

$$P\{\sigma_n = i\} = P\{\sigma_n = i \mid X_0, \ldots, X_n\} = p_i, \qquad 1 \le i \le N,$$

determines a Markov chain $\{X_n\}_{n=0}^{\infty}$ on $\mathbb{R}^2$. Based on this chain is a probabilistic algorithm for generating digitized images, also known as the *random iteration algorithm* or the *chaos game*, which can be described in the following way.

Select any point x_0 in the plane and, independently of it, randomly choose one of the N affine transformations according to the specified probabilities. Apply that transformation, e.g. w_σ, to x_0, and compute a new point $x_1 = w_\sigma(x_0)$. Independently of x_0 and x_1, randomly select another transformation and apply it to x_1, thus obtaining a new point x_2. Iterate the previous steps so that in general, at stage n, a new transformation is chosen independently of all the points that have been previously generated, and x_{n+1} is computed by applying the selected transformation to x_n. The sequence $\{x_n\}_{n=0}^{\infty}$ constitutes a trajectory of the Markov chain $\{X_n\}_{n=0}^{\infty}$. If the chain $\{X_n\}_{n=0}^{\infty}$ is ergodic, in the sense that the empirical distributions of almost every trajectory converge weakly to the unique stationary probability ν of the chain, then, by plotting the points of the sequence $\{x_n\}_{n=0}^{\infty}$ for $n \geq n_0$ sufficiently large, one obtains an approximate image of the support of ν.

Ergodicity of $\{X_n\}_{n=0}^{\infty}$ is key to the implementation of the algorithm, since it guarantees that, for a given set of mappings and associated probabilities, the same image will always be obtained, regardless of the starting point and of the particular sequence of mappings being selected. A sufficient condition for $\{X_n\}_{n=0}^{\infty}$ to be ergodic is that all the transformations be contractive, with respect to some metric on $\mathbb{R}^2$. This condition alone is also sufficient to ensure (see Berger (1989b)) that the support A of ν is bounded, and coincides with the unique, nonempty, compact subset of $\mathbb{R}^2$ satisfying the relation:

$$A = \bigcup_{i=1}^{N} w_i(A). \tag{3.2}$$

We can notice immediately that the characterization of the support of ν provided by Equation (3.2) coincides with the characterization of the attractor of the IFS provided by Equation (2.2). This implies that the support of the unique stationary distribution of the Markov chain $\{X_n\}_{n=0}^{\infty}$ coincides, when the IFS is hyperbolic, with the attractor of the IFS, as defined in the previous chapter.

As noted before, this is a very important characterization of images that are obtainable through this algorithm. It states that every such image can be covered with N affine copies of itself, i.e., by a finite collection of images whose shapes still resemble that of the resulting image. In other words, the image can be regarded as a puzzle, where all the pieces are affinely similar to it, and where overlapping of different pieces is allowed. It is also clear from relation (3.2) that the support of ν is independent of the specific (nonzero) probabilities associated with the transformations in Equation (3.1).

In spite of the inherent simplicity of the algorithm, a great variety of differently shaped images can be generated, even when a small number of transformations is employed (see the images presented in the previous chapter, and Barnsley and Sloan (1988), for some more elaborate examples). This fact, together with the sizable data compression that can be achieved, makes the algorithm particularly attractive. While digitized images are usually stored as large arrays of data in which each entry contains information relative to a corresponding pixel on the screen, the approach described here allows one to regard the coefficients of the mappings and the values of the associated probabilities as an encoding of the image. The algorithm itself should then be viewed as an essential part of the storage, since the image can always be reproduced on the screen by simply running a trajectory of the chain $\{X_n\}_{n=0}^{\infty}$ and successively plotting the points of the sequence as they are generated. The fact that the algorithm is merely based on affine arithmetic makes its actual implementation very fast, so that enormous storage space reduction is attained through a procedure that is, at the same time, extremely efficient and versatile.

The basic algorithm described above can be modified and extended in different ways. A first, simple modification allows generation of full-color images. In addition to running a trajectory of the Markov chain and plotting the points, one can keep track of the frequencies with which pixels on the screen are visited by the trajectory. The image can then be colored by assigning different colors to different frequencies according to some specified color map. Hence, while a black-and-white image represents only the support of the stationary probability ν, a full-color image gives a complete description of the whole distribution, by means of the correspondence between colors and frequencies.

Notice that the particular assignment of probabilities to the transformations comprising a given IFS only affects the coloring of the resulting image, since the shape is uniquely characterized by Equation (3.2). Here are the assignments of probabilities used to generate some of the images presented in this book. The attractors displayed in Figures 2.1, 2.2, and 2.3 were obtained by assigning equal probabilities to all the transformations comprising the corresponding IFSs. The transformations listed on page 22 were assigned respective probabilities $p_1 = 0.5$, $p_2 = 0.168$, $p_3 = 0.166$, and $p_4 = 0.166$, when generating the maple leaf of Figure 2.4, and the transformations listed on page 23 were assigned respective probabilities $p_1 = 0.73$, $p_2 = 0.13$, $p_3 = 0.13$, and $p_4 = 0.01$, when generating the fern displayed in Figure 2.5.

Another possible modification of the algorithm, which will be treated in the sequel, is based on a mixing technique, wherein orbits of different Markov chains are generated simultaneously and merged together according to a probabilistic scheme. Images obtained in this way have shapes which combine features and forms of the images corresponding to the original chains.

Finally, we would like to point out that the performance and efficiency of the basic algorithm can be further improved through parallel implementation, which can be accomplished by running many orbits of the chain simultaneously, starting at different points. The various partial results can then be tallied together in one last step to produce the final image.

3.2.   Stochastic Model

The notion of an IFS with probabilities has been briefly described in Section 3.1 for the case of affine transformations of $\mathbb{R}^2$ into itself and finitely supported distributions on the class of such transformations. In this section, some possible generalizations are presented.

IFSs based on affine transformations of $\mathbb{R}^m$ into $\mathbb{R}^m$ have been considered in the literature. The discussion that follows is mainly based on the approach taken by Berger (1989b). [1] We recall that an affine transformation of $\mathbb{R}^m$ into itself is a mapping of the form $w(x) = Ax + b$, where $A = A(w)$ and $b = b(w)$ denote, respectively, the linear part and the translational part of w. The collection $G = G_m$ of such transformations is finite dimensional, and can be endowed with a metric that makes it a locally compact Hausdorff space. Let $\mathcal{P}(G)$ denote the collection of all Borel probabilities on G. If μ is an element of $\mathcal{P}(G)$ and $\{\mathcal{W}_n\}_{n=0}^{\infty}$ is a sequence of independent random transformations from G with common distribution μ, it is possible to associate with μ a Markov chain $\{X_n\}_{n=0}^{\infty}$ defined by:

$$X_{n+1} = \mathcal{W}_{n+1}(X_n), \qquad n = 0, 1, 2, \ldots \ .$$

A distribution ν belonging to $\mathcal{P}(\mathbb{R}^m)$, the collection of all Borel probabilities on $\mathbb{R}^m$, is said to be μ-*stationary* if, and only if,

$$\mu * \nu = \nu. \tag{3.3}$$

Here, for all Borel subsets B of $\mathbb{R}^m$, the *convolution* $\mu * \nu$ is defined as:

$$\mu * \nu \, (B) = \int \nu(w^{-1} B) \, d\mu(w). \tag{3.4}$$

In other words, if $\mathcal{W}$ is distributed like μ and X is distributed like ν, with $\mathcal{W}$ and X independent, then $\mu * \nu$ is the distribution of $\mathcal{W}(X)$. It is worthwhile noting that, if X_0 is distributed like ν, and ν is μ-stationary, then the chain $\{X_n\}_{n=0}^{\infty}$ associated with μ is stationary.

The main results relative to the stochastic model introduced in this section are outlined in the following theorem, whose proof can be found in Berger (1989b).

Theorem 3.1. *Assume that the probability measure $\mu \in \mathcal{P}(G)$ satisfies the following three conditions:*

$$\int \log^+ \|A(w)\| \, d\mu(w) < \infty, \tag{3.5}$$
$$\int \log^+ |b(w)| \, d\mu(w) < \infty, \tag{3.6}$$
$$\mathbf{E} \log \|A_n \cdots A_1\| < 0 \qquad \textit{for some } n, \tag{3.7}$$

---

[1]Some of the results presented in Berger (1989b) have now also been published in Berger (1993).

where $|\cdot|$ and $\|\cdot\|$ denote compatible vector and matrix norms, respectively, and $\mathcal{A}_k$, for $1 \leq k \leq n$, denotes the linear part of the random affine transformation $\mathcal{W}_k$. Such conditions are sufficient for μ to have the following properties:

(P1) *There exists a unique $\nu \in \mathcal{P}(\mathbb{R}^m)$ which is μ-stationary.*

(P2) *Let $\mu^{(n)}$ denote the n-fold convolution of μ. Then, for any $\nu_0 \in \mathcal{P}(\mathbb{R}^m)$, $\mu^{(n)} * \nu_0$ converges weakly to ν. This property can also be restated in terms of the Markov chain $\{X_n\}_{n=0}^{\infty}$ associated with μ in the following way. For any initial distribution ν_0 on X_0, X_n converges weakly, as n goes to infinity, to X, where X is distributed like ν.*

(P3) *If $\{X_n\}_{n=0}^{\infty}$ is the chain associated with μ, with X_0 having an arbitrary distribution, then, for any bounded continuous function $f : \mathbb{R}^m \mapsto \mathbb{R}$,*

$$\frac{1}{n+1} \sum_{k=0}^{n} f(X_k) \longrightarrow \int f \, d\nu \qquad a.s., \text{ as } n \to \infty.$$

An equivalent statement of this property is that, for almost every trajectory, the empirical distributions $(n+1)^{-1} \sum_{k=0}^{n} \delta_{X_k}$ converge weakly to ν, as n goes to infinity. (By definition, given a point $x \in \mathbb{R}^m$, and for any Borel subset B of $\mathbb{R}^m$, $\delta_x(B)$ equals one or zero, depending on whether B contains x or not.)

The previous theorem constitutes the cornerstone on which the random iteration algorithm for image generation is built. In fact, the theory introduced in Section 3.1 can be viewed as a special case of the present, more general setup. This is easily seen if one sets $m = 2$ and lets μ be *finitely supported*. A finitely supported probability μ, here, is one that assigns measure one to a finite set of elements of G. In other words, for $w_1, \ldots, w_N$ belonging to G and $p_1, \ldots, p_N$ a set of positive weights summing to one, there holds $\mu(\{w_i\}) = p_i$, $1 \leq i \leq N$. Property (P1), then, guarantees the existence of a unique stationary distribution for the Markov chain $\{X_n\}_{n=0}^{\infty}$, and property (P3) provides the convergence result for the empirical distributions of $\{X_n\}_{n=0}^{\infty}$, on which the probabilistic image generation algorithm hinges.

When the measure μ is finitely supported on a set of N affine transformations belonging to G_m and condition (3.7) is satisfied, Equation (3.3) translates into the fundamental invariance condition:

$$\nu(B) = \sum_{i=1}^{N} p_i \, \nu \left(w_i^{-1}(B) \right), \tag{3.8}$$

for every Borel subset B of $\mathbb{R}^m$.

We would like to illustrate the scope of the theorem by reexamining the simple example presented on page 17 in the deterministic context. Consider the IFS with probabilities $\{\mathbb{R}; \, w_1(x) = (1/2)x,$ $w_2(x) = (1/2)x + 1/2; \, p_1 = p_2 = 1/2\}$, where $\mathbb{R}$ is endowed with the Euclidean metric. It can be easily verified that, in this case, the hypotheses of Theorem 3.1 are satisfied, so that, by property (P1), the Markov chain $\{X_n\}_{n=0}^{\infty}$ associated with this IFS has a unique stationary distribution on $\mathbb{R}$. We have seen in the previous chapter that the attractor of this IFS coincides with the interval $[0,1]$. As we have observed in the previous section (for the similar case of IFSs on $\mathbb{R}^2$), since the IFS is hyperbolic, such an interval must also be the support of the stationary distribution for $\{X_n\}_{n=0}^{\infty}$. We shall now see that the stationary distribution ν must be uniform on $[0,1]$.

In order to show that this is the case, it is enough to verify that the invariance condition of Equation (3.8) is satisfied for any set of the form $(-\infty, x]$, with $x \in \mathbb{R}$. In this particular case, such a condition can be written as:

$$
\begin{aligned}
\nu \left((-\infty, x] \right) &= \sum_{i=1}^{2} p_i \, \nu \left(w_i^{-1}(-\infty, x] \right) \\
&= \frac{1}{2} \nu \left((-\infty, 2x] \right) + \frac{1}{2} \nu \left((-\infty, 2x - 1] \right).
\end{aligned}
$$

Now, if $x < 0$, $\nu \left((-\infty, x] \right) = 0$ and the right-hand side equals $(1/2) \cdot 0 + (1/2) \cdot 0 = 0$. If $0 \le x < (1/2)$, $\nu \left((-\infty, x] \right) = x$ and the right-hand side equals $(1/2) \cdot 2x + (1/2) \cdot 0 = x$. If $(1/2) \le x < 1$, $\nu \left((-\infty, x] \right) = x$ and the right-hand side equals $(1/2) \cdot 1 + (1/2) \cdot (2x - 1) = x$. Finally, if $x \ge 1$, $\nu \left((-\infty, x] \right) = 1$ and the right-hand side equals $(1/2) \cdot 1 + (1/2) \cdot 1 = 1$. Hence, the invariance condition being satisfied, the uniform distribution on $[0, 1]$ is indeed the stationary distribution for the Markov chain associated with the given IFS.

3.3.   Self-Covering Property

It is shown in Berger (1989b) that, under the theoretical setup of the previous section, the support A of a μ-stationary probability ν belonging to $\mathcal{P}(\mathbb{R}^m)$, i.e., the intersection of all closed sets of ν-measure one, satisfies the following, important, self-covering property:

$$A = \overline{\bigcup_{w \in G'} w(A)}, \tag{3.9}$$

where G' is the support of μ and the overlying bar denotes closure.

When the measure μ is finitely supported, in the sense defined on page 46, the self-covering condition (3.9) translates into:

$$A = \bigcup_{i=1}^{N} \overline{w_i(A)}. \tag{3.10}$$

One can easily verify that the first two technical conditions (3.5) and (3.6) are automatically satisfied when μ is finitely supported. Hence, if μ is chosen in such a way that condition (3.7) is also satisfied, then property (P3) guarantees that the algorithm described in Section 3.1 will produce, with probability one, the image corresponding to the unique μ-stationary probability ν. As noted in Section 3.1 the black-and-white image will represent the support of ν and satisfy the self-covering property (3.10).

In view of what has just been said, it is important to be able to determine when there exists an assignment of probabilities p_i such that the measure μ satisfies condition (3.7). As noted in Berger (1989b), this happens, for instance, if at least one of the transformations is strictly contractive, provided that the probability associated with such a transformation is chosen sufficiently close to one. As it has already been stated in Section 3.1, when all the N affine transformations are strict contractions, something more can be said; namely, that the support A of the μ-stationary probability ν is the *unique* nonempty *compact* set which satisfies the self-covering condition (3.2), and coincides with the attractor of the IFS defined in Chapter 2. Notice that, all the transformations being continuous and A being compact, the closure sign does not appear on the right-hand side of condition (3.2).

If one wants to guarantee the existence of a *unique* compact self-covering set, the assumption that all transformations are strict contractions cannot, in general, be weakened (see Berger and Amit (1987)). The case where all N mappings coincide with the identity provides a straightforward example of a situation where any compact set is self-covering, although all transformations are (not strictly) contractive. Even in the case of strict contractions, uniqueness does not necessarily hold, if one's attention is not restricted to *compact* sets. For instance, the origin $\{O\}$ is the unique compact self-covering set for the IFS consisting of only the transformation $w = aI$, where $0 < a < 1$ and I denotes the identity mapping, but any ray emanating from O also satisfies condition (3.2).

3.4.   Invariance Condition and Encoding

As pointed out in Section 3.1, the stationary distribution of the Markov chain associated with an IFS with probabilities on the plane can be naturally identified with a color image. Practical applications would then require, given an arbitrary color image, that one be able to solve the inverse problem of determining an IFS with probabilities, whose associated image coincides with (or at least approximates) the given image.

The nature of this problem is completely similar to that of encoding black-and-white images by means of deterministic IFSs, as described in Section 2.5. The comments made there apply, *mutatis mutandis,* to the case we are now considering, with the understanding that the additional information conveyed by the coloring can be used to try to estimate, besides the transformations themselves, the probabilities associated with them.

In this section, we intend to review a couple of methods of addressing the solution of the inverse problem that have been proposed in the literature, and are based on property (P1) of page 46. For the sake of simplicity, we shall limit our discussion to the hypothetical situation in which the image under consideration is known to be associated with an IFS with probabilities, consisting of N affine transformations of the plane into itself. The unknown quantities are, in this case, the $6N$ parameters that define the transformations and the N associated probabilities. Notice, incidentally, that as a simple consequence of the invariance condition (3.8), once the transformations are known, determination of the probabilities from the color is a linear problem.

The first method that we are going to describe, known as the *method of moments,* exploits the linear structure of Equation (3.8) and of the affine transformations w_i to obtain formulae for the moments of ν. If $X = (X_1, X_2)^T$ is distributed like ν, then, denoting by w_{1i} and w_{2i} the first and second components of w_i, it follows from Equation (3.8) that:

$$\mathbf{E}\left[X_1^k X_2^l\right] = \sum_{i=1}^{N} p_i \, \mathbf{E}\left[(w_{1i}(X))^k (w_{2i}(X))^l\right]. \qquad (3.11)$$

Because of the linearity of the operator $\mathbf{E}$, the right-hand side of Equation (3.11) can be expanded as a sum of various moments of X, weighted by polynomial functions of the coefficients of the transformations w_i and of the probabilities p_i. It then becomes only a matter of simple algebra to rewrite the moments as rational functions of the coefficients of the transformations and of the associated probabilities.

Therefore, when the distribution of X is given, in the form of a full-color image for example, one can numerically evaluate as many moments as needed and try to determine the coefficients of the mappings w_i and the associated probabilities p_i, by solving a system of polynomial equations. The solution of such a system constitutes, in general, a nontrivial numerical problem. This is because of the large number of unknowns that are usually involved, and because of the existence of solutions that do not correspond to an ergodic Markov chain (the trivial solution $w_i = I$, $i = 1, \dots, N$, being such a case). A practical example of how this method can be successfully employed can be found in Barnsley and Demko (1985).

The second method that we are now going to describe is based on the *Radon transform*. An examination of the invariance condition (3.8) suggests identification of a class of Borel subsets of $\mathbb{R}^2$ whose preimages under affine transformations still belong to the same class. As it turns out, half-planes have this property, i.e., the preimage of a half-plane under a nondegenerate affine transformation is still a half-plane. To make this more precise, given any unit vector $\lambda = (\lambda_1, \lambda_2)^T \in \mathbb{R}^2$ and any real number z, define:

$$H(\lambda, z) = \left\{ x \in \mathbb{R}^2 : x^T \lambda \leq z \right\}.$$

$H(\lambda, z)$ is the half-plane underneath the line of equation $x^T \lambda = z$, i.e., the line with normal direction λ and distance from the origin $|z|$. It is easily seen that the preimage of $H(\lambda, z)$ under the affine transformation $w(x) = Ax + b$ is given by the half-plane

$$w^{-1} H(\lambda, z) = H\left(\frac{A^T \lambda}{|A^T \lambda|}, \frac{z - b^T \lambda}{|A^T \lambda|} \right).$$

When $A^T \lambda = 0$, then $w^{-1} H(\lambda, z)$ is either empty or it coincides with the whole $\mathbb{R}^2$.

Now, letting $R(\lambda, z) = \nu \left\{ H(\lambda, z) \right\}$, the invariance condition (3.8) can be rewritten for the case of half-planes as:

$$R(\lambda, z) = \sum_{i=1}^N p_i \, R\left(\frac{A_i^T \lambda}{|A_i^T \lambda|}, \frac{z - b_i^T \lambda}{|A_i^T \lambda|} \right). \tag{3.12}$$

The function $R(\lambda, z)$ defined above is known as the Radon transform of the probability ν. For any fixed λ, when regarded as a function of z, it constitutes a univariate c.d.f. It can be shown (see Deans (1983) for further details) that the specification of R is equivalent to the specification of the bivariate c.d.f. of ν, in the sense that any of them can be obtained from the other.

Given a full-color image, it is possible to evaluate the function R and solve Equation (3.12) for the coefficients of the transformations w_i and the associated weights p_i. Once again, the numerical difficulties that the solution of this problem poses are not, in general, negligible. A clever example of the use of the Radon transform towards the solution of an inverse problem with available partial information is given in Berger and Amit (1987).

3.5. Mixing Algorithm

More elaborate images than those obtainable by means of the basic generation algorithm of Section 3.1 can be generated by merging together, by means of a probabilistic mixing technique, orbits of several Markov chains corresponding to different IFSs with probabilities. Here, we shall only consider a simple case of mixing, based on two Markov chains, which illustrates, nevertheless, all the important features of the general technique. Proofs of the results that we shall outline, together with a development of the theory in a more general setting, can be found in Barnsley et al. (1988).

The probabilistic setting and the notation that we use coincide with those employed in Section 3.2, when discussing the case of a single Markov chain. Here, however, we consider two independent sequences $\{V_n\}_{n=0}^{\infty}$ and $\{W_n\}_{n=0}^{\infty}$ of independent, identically distributed, random, affine transformations, with respective distributions μ_V and μ_W, and proceed to mix together the two associated Markov chains.

The process $\{X_n\}_{n=0}^{\infty}$ associated with the first sequence of random transformations is generated according to the same scheme outlined in Section 3.2, by means of the recursion:

$$X_{n+1} = V_{n+1}(X_n), \qquad n = 0, 1, 2, \ldots \ . \tag{3.13}$$

In order to generate the second process $\{Y_n\}_{n=0}^{\infty}$, let p, with $0 < p < 1$, and a deterministic affine transformation u be given, and define recursively:

$$Y_{n+1} = \begin{cases} u(X_n) & \text{with probability } p \\ W_{n+1}(Y_n) & \text{with probability } 1-p \end{cases} \qquad n = 0, 1, 2, \ldots,$$

$$\tag{3.14}$$

where the choice is made independently of $\{V_n\}_{n=0}^{\infty}$ and $\{W_n\}_{n=0}^{\infty}$.

It is shown in Barnsley et al. (1988) that the process $\{(X_n, Y_n)\}_{n=0}^{\infty}$ is a Markov chain on $\mathbb{R}^m \times \mathbb{R}^m$ which is asymptotically stationary, provided the random sequence of transformations $\{V_n\}_{n=0}^{\infty}$ satisfies the assumptions of Theorem 3.1. More details on the form of the joint limiting distribution of the process are provided in the aforementioned paper.

What is important for the purpose of image generation is that, whenever $\mu_{\mathcal{W}}$ is finitely supported, i.e., when $\mu_{\mathcal{W}}(\{w_i\}) = p_i$, for $1 \leq i \leq N$, with $\sum_{i=1}^{N} p_i = 1$, the support C_Y of the limiting distribution of the process $\{Y_n\}_{n=0}^{\infty}$ satisfies the condition:

$$C_Y = \overline{u\,(C_X)} \bigcup \left(\bigcup_{i=1}^{N} \overline{w_i\,(C_Y)} \right), \qquad (3.15)$$

where C_X is the support of the unique invariant distribution ν for the process $\{X_n\}_{n=0}^{\infty}$. Furthermore, if $\mu_{\mathcal{V}}$ is also finitely supported, and all transformations are strictly contractive with respect to some metric, the closure signs can be removed from Equation (3.15), and C_Y becomes the unique, nonempty, compact subset of $\mathbb{R}^m$ such that:

$$C_Y = u\,(C_X) \bigcup \left(\bigcup_{i=1}^{N} w_i\,(C_Y) \right). \qquad (3.16)$$

Equation (3.16) should be compared to Equation (2.3) of page 29, which provides the covering characterization for the attractor of an IFS with condensation. The image under u of the support of the invariant distribution ν coincides with the condensation set C, and C_Y is the attractor of the IFS with condensation consisting of $w_1, \ldots, w_N$, and of the additional transformation that maps every nonempty, compact subset of $\mathbb{R}^m$ into $u\,(C_X)$.

In the case when $\mu_{\mathcal{V}}$ and $\mu_{\mathcal{W}}$ are both finitely supported on the class of affine transformations of the Euclidean plane into itself, Equations 3.13 and 3.14 provide the basis for an image generation algorithm, known as the *mixing algorithm*. A suggestive way of thinking about the procedure, as explained in Barnsley et al. (1988), requires imagining that a sample path of $\{X_n\}_{n=0}^{\infty}$ is generated on an upper screen according to Equation (3.13), and that a sample path of $\{Y_n\}_{n=0}^{\infty}$ is generated, synchronously, on a lower screen according to Equation (3.14).

While the orbit of $\{X_n\}_{n=0}^{\infty}$ evolves according to the random sequence $\{\mathcal{V}_n\}_{n=0}^{\infty}$, independently of what happens on the lower screen, the orbit of $\{Y_n\}_{n=0}^{\infty}$ on the lower screen either evolves according to the random sequence $\{\mathcal{W}_n\}_{n=0}^{\infty}$ with probability $1-p$, or jumps, with probability p, to a point which is the image under u of the current state in which the orbit of $\{X_n\}_{n=0}^{\infty}$ happens to be. It is intuitively appealing to think of such a point as if it were dropped from the upper onto the lower screen. Only the lower screen is eventually plotted, and the resulting image will represent the support of the limiting distribution of the process $\{Y_n\}_{n=0}^{\infty}$.

As in the case of the basic image generation algorithm, color can be added according to the usual frequency-based scheme. When all the transformations involved are strict contractions, the resulting image C_Y satisfies Equation (3.16), and it is therefore the attractor of a hyperbolic IFS with condensation. Hence, as mentioned in Section 2.4, the algorithm provides a speedy and efficient way of generating the attractor of an IFS with condensation, whenever the condensation set can be regarded as the image, under an affine transformation u, of the attractor of another IFS.

As an example, consider the image of Figure 3.1 and Color Plate 5. The condensation set is given by the fern stemming from the tip of the maple leaf. It is obtained by generating a fern on the upper screen by means of the basic algorithm, using the transformations listed on page 23 with the respective, associated probabilities listed on page 44, and mapping it into the lower screen through the transformation:

$$u\left((x_1, x_2)^T\right) = \begin{pmatrix} 0.33 & 0 \\ 0 & 0.33 \end{pmatrix} \begin{pmatrix} x_1 \\ x_2 \end{pmatrix} + \begin{pmatrix} 0.339 \\ 0.773 \end{pmatrix}.$$

The mixing probability p is chosen to equal 0.05. The transformations characterizing the process generated on the lower screen are those that, when used in conjunction with the basic algorithm, generate the maple leaf of Figure 2.4. They are listed on page 22, and their associated probabilities are listed on page 44. Observe that, even in the mixing algorithm, the assignment of probabilities to the different transformations affects the coloring but not the shape of the resulting image.

3.6.   Another Look at the Stochastic Model

Probabilistic IFS theory has been presented in Section 3.2 for the case in which all transformations involved are affine mappings of

Figure 3.1. Mixing of maple leaf and black spleenwort fern.

$\mathbb{R}^m$ into itself. The results that have been stated there, and in particular the existence of a unique invariant distribution associated with a given IFS with probabilities, can be proved by means of a reversibility phenomenon concerning the Markov chain $\{X_n\}_{n=0}^{\infty}$, and ergodic theory arguments (see Berger (1989b) for details).

The reason we decided to restrict our attention to the case of Euclidean spaces, and affine transformations on such spaces, is solely due to the fact that they provide the most natural context, and manageable tools, for applying IFS theory to the encoding and decoding of digitized images. However, IFSs defined on other spaces, and by means of different transformations, have also been considered in the literature.

In this section, we would like to look at another possible approach to the development of probabilistic IFS theory on a generic, compact metric space (X, d), along the lines followed in Chapter 9 of Barnsley (1988). Our interest in such an approach is not motivated by the fact that non-Euclidean spaces are considered, but rather by the fact that, in this framework, the unique invariant probability associated with a given IFS with probabilities can be regarded as the fixed point of a certain contractive transformation on the space of Borel probabilities on X, thereby providing a nice parallel with the development of deterministic IFS theory presented in Chapter 2.

Let (X, d) then denote a compact metric space, and let $\mathcal{P}(X)$ denote the class of Borel probabilities on X. In order to be able to apply the contraction mapping theorem to a function defined on $\mathcal{P}(X)$, $\mathcal{P}(X)$ has to be made into a metric space. This can be done using the so-called *Hutchinson* metric, defined by:

$$d_H(\lambda, \mu) = \sup \left\{ \left. \int_X f \, d\lambda - \int_X f \, d\mu \right| f \in LC_1 \right\}, \quad \forall \, \lambda, \mu \in \mathcal{P}(X),$$

where

$$LC_1 = \left\{ f : X \longmapsto \mathbb{R} \, \middle| \, |f(x) - f(y)| \leq d(x, y), \, \forall \, x, y \in X \right\}$$

denotes the class of Lipschitz-continuous functions from X into $\mathbb{R}$ with Lipschitz constant 1. It is shown in Hutchinson (1981) that the space $\mathcal{P}(X)$ of Borel probabilities on X, endowed with the Hutchinson metric, is a compact metric space.

Suppose now that a hyperbolic IFS with probabilities $\{X; w_1, \ldots, w_N; p_1, \ldots, p_N\}$ on the compact metric space (X, d) is given. It is possible to associate with the given IFS a function $M : \mathcal{P}(X) \longmapsto \mathcal{P}(X)$, called the *Markov operator*, defined by:

$$M(\lambda)(B) = \sum_{i=1}^{N} p_i \lambda \left(w_i^{-1}(B) \right), \quad \forall \lambda \in \mathcal{P}(X), \text{ and } B \in \mathcal{B}(X),$$

$$(3.17)$$

where $\mathcal{B}(X)$ denotes the class of Borel subsets of X.

The main result concerning the Markov operator defined above is given in the following theorem, due to Hutchinson (1981), which is also proved in Barnsley (1988).

Theorem 3.2. *Let* $\{X; w_1, \ldots, w_N; p_1, \ldots, p_N\}$ *be a hyperbolic IFS with probabilities on a compact metric space* (X, d). *Let* s *be a con-*

tractivity factor for the IFS. Then, the Markov operator defined by Equation (3.17) is a contraction on $\mathcal{P}(X)$ *endowed with the Hutchinson metric, with contractivity factor* s. *Hence, by the contraction mapping theorem, M has a unique fixed point* ν, *which is called the invariant (or stationary) probability of the given IFS with probabilities. Its support coincides with the attractor of the deterministic IFS* $\{X; w_1, \ldots, w_N\}$.

In light of the previous theorem we have that, at least when the space (X, d) is compact and all transformations are strictly contractive, the following parallel holds: the attractor of a deterministic IFS is the unique fixed point of the contractive transformation W on the space $(\mathcal{H}(X), h)$ defined by Equation (2.1) on page 16, and, likewise, the invariant distribution associated with an IFS with probabilities is the unique fixed point of the contractive transformation M on $(\mathcal{P}(X), d_H)$ defined by Equation (3.17).

It would seem, at first, that the requirement that (X, d) be compact, precludes the possibility of applying Theorem 3.2 to the case of IFSs on Euclidean spaces. However, it is shown in Barnsley (1988) that, given a hyperbolic IFS on a complete metric space (X, d), there always exists a compact subset K of X such that $W(K) \subset K$, so that $w_1, \ldots, w_N$ can be thought of as mappings from K into K, rather than from X into X.

As in the deterministic case, there is also a probabilistic equivalent of the collage theorem (see Barnsley (1988) for details), which can be used, in a manner completely similar to the one described in Section 2.5, to attack the inverse problem. The practical application of such an approach is hampered, however, by the computational difficulties encountered when evaluating the Hutchinson distance between probability measures.

Suppose now that $\{X; w_1, \ldots, w_N; p_1, \ldots, p_N\}$ is a hyperbolic IFS with probabilities on a compact metric space (X, d) and that $\{X_n\}_{n=0}^{\infty}$ is the associated Markov chain defined recursively by Equation (3.1) on page 41. It is easy to verify that if X_n is distributed like ν_n, then X_{n+1} is distributed like $M(\nu_n)$. It then follows from the convergence part of the contraction mapping theorem, from Theorem 3.2, and from the fact that the Hutchinson metric induces the topology of weak convergence that, for any initial distribution ν_0 on X_0, X_n converges weakly, as n goes to infinity, to a random variable distributed like ν, where ν is the unique invariant probability associated with the given IFS. Hence, properties (P1) and (P2) of Theorem 3.1 still hold in this setting. It is actually possible to show that $\{X_n\}_{n=0}^{\infty}$ has a unique limiting invariant distribution ν even when the metric space (X, d) is only locally compact (Euclidean, for instance), and the probabilities $p_1, \ldots, p_N$ are place-dependent, provided they satisfy some additional conditions (see Elton (1987) for details). The following theorem, due to Elton, is similar to property (P3) of Theorem 3.1, and is stated here in the form that we shall need in the following chapter. Its proof can also be found in Elton (1987).

Theorem 3.3. *Let $\{X; w_1, \ldots, w_N; p_1, \ldots, p_N\}$ be a hyperbolic IFS with probabilities on a locally compact metric space (X, d), and let $\{X_n\}_{n=0}^{\infty}$ be its associated Markov chain with unique, invariant, limiting distribution ν. Then, for any starting point $x_0 \in X$, and any bounded, continuous function $f : X \mapsto \mathbb{R}$,*

$$\frac{1}{n+1} \sum_{k=0}^{n} f(x_k) \longrightarrow \int_X f(x) \, d\nu(x), \qquad \text{as } n \to \infty,$$

for almost every orbit $\{x_n\}_{n=0}^{\infty}$ of $\{X_n\}_{n=0}^{\infty}$.

3.7. Other Applications of the Invariance Condition

Although in this book we are mainly concerned with representing digitized images through IFS encoding, this is by no means the only possible application of the theory. In this section, we shall see how the invariance condition stated in Equation (3.8) can be usefully utilized in other settings.

As a first example, let X_1, X_2 be a random sample of size two from a uniform distribution on $[0, 1]$. Let $L = X_{(2)} - X_{(1)}$ denote the difference between the maximum and the minimum of the sample. In other words, $L = |X_1 - X_2|$ denotes the length of a segment whose endpoints are independently drawn at random from the unit interval. Suppose we want to determine the kth moments $\mathbf{E}L^k$ of L, for $k = 1, 2, \ldots$. The usual approach would require one to determine first the distribution of the random variable L, and then compute the various moments by integration.

It is actually not difficult to show (see, for instance, David (1981)) that L is distributed like a Beta $(1, 2)$ random variable, where we say that a random variable X has a Beta (a, b) distribution if and only if its density function with respect to Lebesgue measure on the real line is given by:

$$f_X(x) = \frac{1}{B(a, b)} x^{a-1} (1 - x)^{b-1} I_{[0,1]}(x), \qquad \text{with } a > 0 \text{ and } b > 0,$$

where the normalizing factor $B(a, b)$ is known as the Beta function, and $I_{[0,1]}$ denotes the indicator of the set $[0, 1]$. By integrating x^k against $f_X(x)\, dx$, for $a = 1$ and $b = 2$, one obtains:

$$\mathbf{E}L^k = \frac{2}{(k + 1)(k + 2)}, \qquad \text{for } k = 1, 2, \ldots . \tag{3.18}$$

We shall now see how the same result can be obtained, without having to determine the distribution of L, by exploiting the invariance condition for the invariant distribution of a suitable IFS with probabilities. Recall, from Section 3.2, that the unique invariant distribution for the IFS with probabilities $\{\mathbb{R}; w_1(x) = (1/2)x, w_2(x) = (1/2)x + 1/2; p_1 = p_2 = 1/2\}$, is uniform on $[0, 1]$, and coincides with the unique limiting distribution of the Markov chain associated with the IFS.

Consider now two chains $\{X_n^1\}_{n=0}^\infty$ and $\{X_n^2\}_{n=0}^\infty$ associated with the given IFS, independent of one another, and construct the new process $\{L_n = |X_n^1 - X_n^2|\}_{n=0}^\infty$. We shall see, in a moment, that $\{L_n\}_{n=0}^\infty$ can be viewed as a Markov chain associated with a hyperbolic IFS with probabilities on the real line. Because of the way the process has been defined, it is then clear that its unique, invariant, limiting distribution must coincide with the distribution of L.

In order to see that $\{L_n\}_{n=0}^\infty$ is indeed a Markov chain, consider first the case when $X_n^2 \geq X_n^1$. Then $L_n = X_n^2 - X_n^1$, and, since for each of the two processes the transformation that is applied at time n is chosen independently with probability $1/2$, we have:

$$L_{n+1} \left| X_n^2 \geq X_n^1 \right. =$$

$$\begin{cases} w_1(X_n^2) - w_1(X_n^1) = \frac{1}{2}X_n^2 - \frac{1}{2}X_n^1 = \frac{1}{2}L_n & \text{w.p. } 1/4, \\ w_2(X_n^2) - w_1(X_n^1) = \frac{1}{2}X_n^2 + \frac{1}{2} - \frac{1}{2}X_n^1 = \frac{1}{2}L_n + \frac{1}{2} & \text{w.p. } 1/4, \\ w_2(X_n^2) - w_2(X_n^1) = \frac{1}{2}X_n^2 + \frac{1}{2} - \frac{1}{2}X_n^1 - \frac{1}{2} = \frac{1}{2}L_n & \text{w.p. } 1/4, \\ w_2(X_n^1) - w_1(X_n^2) = \frac{1}{2}X_n^1 + \frac{1}{2} - \frac{1}{2}X_n^2 = -\frac{1}{2}L_n + \frac{1}{2} & \text{w.p. } 1/4, \end{cases}$$

or

$$L_{n+1} \Big| X_n^2 \ge X_n^1 = \begin{cases} \frac{1}{2}L_n & \text{w.p. } 1/2, \\ \frac{1}{2}L_n + \frac{1}{2} & \text{w.p. } 1/4, \\ -\frac{1}{2}L_n + \frac{1}{2} & \text{w.p. } 1/4. \end{cases}$$

Similarly, if $X_n^1 > X_n^2$, $L_n = X_n^1 - X_n^2$, and we have:

$$L_{n+1} \Big| X_n^1 \ge X_n^2 =$$

$$\begin{cases} w_1(X_n^1) - w_1(X_n^1) = \frac{1}{2}X_n^1 - \frac{1}{2}X_n^2 = \frac{1}{2}L_n & \text{w.p. } 1/4, \\ w_2(X_n^1) - w_1(X_n^2) = \frac{1}{2}X_n^1 + \frac{1}{2} - \frac{1}{2}X_n^2 = \frac{1}{2}L_n + \frac{1}{2} & \text{w.p. } 1/4, \\ w_2(X_n^1) - w_2(X_n^2) = \frac{1}{2}X_n^1 + \frac{1}{2} - \frac{1}{2}X_n^2 - \frac{1}{2} = \frac{1}{2}L_n & \text{w.p. } 1/4, \\ w_2(X_n^2) - w_1(X_n^1) = \frac{1}{2}X_n^2 + \frac{1}{2} - \frac{1}{2}X_n^1 = -\frac{1}{2}L_n + \frac{1}{2} & \text{w.p. } 1/4, \end{cases}$$

or

$$L_{n+1} \Big| X_n^1 \ge X_n^2 = \begin{cases} \frac{1}{2}L_n & \text{w.p. } 1/2, \\ \frac{1}{2}L_n + \frac{1}{2} & \text{w.p. } 1/4, \\ -\frac{1}{2}L_n + \frac{1}{2} & \text{w.p. } 1/4. \end{cases}$$

It is clear from the above that conditioning on which of X_n^1 and X_n^2 is the lower endpoint and the upper endpoint of the interval is immaterial, and that the transition probabilities of the process depend only on its current state. Thus, $\{L_n\}_{n=0}^\infty$ is a Markov chain. It also follows that $\{L_n\}_{n=0}^\infty$ coincides with the Markov chain associated with the hyperbolic IFS with probabilities $\{\mathbb{R}; v_1(x) = (1/2)x, v_2(x) = (1/2)x + 1/2, v_3(x) = -(1/2)x + 1/2; q_1 = 1/2, q_2 = q_3 = 1/4\}$.

As we have already remarked, the unique invariant distribution for such an IFS must coincide with the distribution of L. We shall make use of this fact to prove by induction that formula (3.18) holds. For $k = 1$, applying the invariance condition, we have:

$$\begin{aligned} \mathbf{E}L &= \sum_{j=1}^3 q_i \, \mathbf{E}(v_j(L)) \\ &= \frac{1}{2}\mathbf{E}\left(\frac{1}{2}L\right) + \frac{1}{4}\mathbf{E}\left(\frac{1}{2}L + \frac{1}{2}\right) + \frac{1}{4}\mathbf{E}\left(-\frac{1}{2}L + \frac{1}{2}\right) \end{aligned}$$

$$= \frac{1}{4}\mathbf{E}L + \frac{1}{8}\mathbf{E}L + \frac{1}{8} - \frac{1}{8}\mathbf{E}L + \frac{1}{8}$$

$$= \frac{1}{4}\mathbf{E}L + \frac{1}{4},$$

which implies that $\mathbf{E}L = 1/3$, so that formula (3.18) is true.

Assuming it true for all positive integers less than, or equal to, $k - 1$, and applying the invariance condition once again, we have:

$$\mathbf{E}L^k = \sum_{j=1}^{3} q_i \, \mathbf{E}(v_j(L))^k$$

$$= \frac{1}{2}\mathbf{E}\left(\frac{1}{2}L\right)^k + \frac{1}{4}\mathbf{E}\left(\frac{1}{2}L + \frac{1}{2}\right)^k + \frac{1}{4}\mathbf{E}\left(-\frac{1}{2}L + \frac{1}{2}\right)^k$$

$$= \frac{1}{2^{k+1}}\mathbf{E}L^k + \frac{1}{2^{k+2}}\mathbf{E}(L+1)^k + \frac{1}{2^{k+2}}\mathbf{E}(1-L)^k$$

$$= \frac{1}{2^{k+1}}\mathbf{E}L^k + \frac{1}{2^{k+2}}\sum_{j=0}^{k}\binom{k}{j}\mathbf{E}L^j + \frac{1}{2^{k+2}}\sum_{j=0}^{k}(-1)^j\binom{k}{j}\mathbf{E}L^j.$$

We now need to distinguish between the case when k is even and the case when k is odd. We shall only work out the details of the calculations for the first case, since the second is very similar. Assuming, then, that k is even, we have:

$$\mathbf{E}L^k = \frac{1}{2^{k+1}}\mathbf{E}L^k + \frac{2}{2^{k+2}}\mathbf{E}L^k + \frac{1}{2^{k+2}}\sum_{j=0}^{k-1}\binom{k}{j}\mathbf{E}L^j +$$

$$\frac{1}{2^{k+2}}\sum_{j=0}^{k-1}(-1)^j\binom{k}{j}\mathbf{E}L^j.$$

Thus, collecting terms and using the induction hypothesis, we have:

$$\mathbf{E}L^k = \left(\frac{2^k}{2^k-1}\right)\left(\frac{1}{2^{k+2}}\sum_{j=0}^{k-1}\binom{k}{j}\frac{2}{(j+1)(j+2)} +\right.$$

$$\left.\frac{1}{2^{k+2}}\sum_{j=0}^{k-1}(-1)^j\binom{k}{j}\frac{2}{(j+1)(j+2)}\right)$$

$$= \left(\frac{1}{2(2^k-1)}\right)\left(\sum_{j=0}^{k-1}\binom{k}{j}\frac{1}{(j+1)(j+2)} +\right.$$

$$\sum_{j=0}^{k-1}(-1)^j \binom{k}{j} \frac{1}{(j+1)(j+2)}\bigg).$$

Now,

$$\sum_{j=0}^{k-1}\binom{k}{j}\frac{1}{(j+1)(j+2)} =$$

$$= \sum_{j=0}^{k-1}\left(\frac{k!}{j!(k-j)!}\right)\left(\frac{1}{(j+1)(j+2)}\right)$$

$$= \sum_{j=0}^{k-1}\frac{k!}{(j+2)!(k-j)!}$$

$$= \frac{1}{(k+1)(k+2)}\sum_{j=0}^{k-1}\binom{k+2}{j+2}$$

$$= \frac{1}{(k+1)(k+2)}\sum_{j=2}^{k+1}\binom{k+2}{j}$$

$$= \left(\frac{1}{(k+1)(k+2)}\right)\left(\sum_{j=0}^{k+2}\binom{k+2}{j} - 1 - 1 - (k+2)\right)$$

$$= \frac{2^{k+2} - 2^2 - k}{(k+1)(k+2)}.$$

Similarly,

$$\sum_{j=0}^{k-1}(-1)^j\binom{k}{j}\frac{1}{(j+1)(j+2)} =$$

$$= \sum_{j=0}^{k-1}(-1)^j\left(\frac{k!}{j!(k-j)!}\right)\left(\frac{1}{(j+1)(j+2)}\right)$$

$$= \sum_{j=0}^{k-1}(-1)^j\left(\frac{k!}{(j+2)!(k-j)!}\right)$$

$$= \frac{1}{(k+1)(k+2)}\sum_{j=0}^{k-1}(-1)^j\binom{k+2}{j+2}$$

$$= \frac{1}{(k+1)(k+2)} \sum_{j=2}^{k+1} (-1)^j \binom{k+2}{j}$$

$$= \left(\frac{1}{(k+1)(k+2)} \right) \left(\sum_{j=0}^{k+2} (-1)^j \binom{k+2}{j} - 1 - 1 + k + 2 \right)$$

$$= \frac{k}{(k+1)(k+2)}.$$

Hence, substituting these last two expressions into the formula for $\mathbf{E}L^k$, we have:

$$\mathbf{E}L^k = \left(\frac{1}{2(2^k - 1)} \right) \left(\frac{2^{k+2} - 2^2 - k}{(k+1)(k+2)} + \frac{k}{(k+1)(k+2)} \right)$$

$$= \frac{2}{(k+1)(k+2)},$$

so that formula (3.18) also holds for k, which completes the proof by induction (the case when k is odd, as we have already said, can be dealt with in a similar manner).

Notice that, if we had not known formula (3.18) in advance, we would have been unable to prove its validity by induction. However, we still could have used the invariance condition to compute the various moments of L recursively.

Consider now the family of IFSs with probabilities $\{\mathbb{R}; w_1(x) = ax, \ w_2(x) = ax + (1 - a); \ p_1 = p, p_2 = 1 - p\}$, with $0 \le a \le 1/2$, and $0 < p < 1$. For any a in the specified range, the corresponding IFS is hyperbolic, and therefore has a unique, invariant distribution, whose support can be shown to be contained in $[0, 1]$. Notice that, for $a = 1/2$ and $p = 1/2$, we obtain the IFS of the previous example. For any value of a strictly less than $1/2$, the support of the unique, invariant distribution is totally disconnected (see Hutchinson (1981)), and for $a = 1/3$, we obtain the Cantor set described in Section 2.3. In all such cases, the choice $p_1 = p_2 = 1/2$ produces an invariant distribution that can be thought of as "uniform" over the attractor of the IFS.

Suppose now that, for a given $a \in [0, 1/2]$ and a given $p \in [0, 1]$, X_1, X_2 is a random sample from the unique, invariant distribution for the IFS, and consider the problem of determining the first moment of the random variable $L = X_{(2)} - X_{(1)} = |X_1 - X_2|$. As we did before, we can consider two independent Markov chains $\{X_n^1\}_{n=0}^{\infty}$ and $\{X_n^2\}_{n=0}^{\infty}$ associated with the given IFS, and construct the new process $\{L_n = |X_n^1 - X_n^2|\}_{n=0}^{\infty}$. Using arguments similar to those employed in the previous example, we can show that $\{L_n\}_{n=0}^{\infty}$ can be regarded as a Markov chain associated with the hyperbolic IFS with probabilities $\{\mathbb{R}; v_1(x) = ax, v_2(x) = ax + (1 - a), v_3(x) = -ax + (1 - a); p_1 = p^2 + (1 - p)^2, p_2 = p_3 = p(1 - p)\}$. Since $\{X_n^1\}_{n=0}^{\infty}$ and $\{X_n^2\}_{n=0}^{\infty}$ are independent and have, as their limiting distribution, the common distribution of X_1 and X_2, the unique, invariant, limiting distribution for $\{L_n\}_{n=0}^{\infty}$ must coincide with the distribution of L.

Applying the invariance condition, we then have:

$$
\begin{aligned}
\mathbf{E}L &= \sum_{j=1}^{3} q_i\, \mathbf{E}(v_i(L)) \\
&= \left(p^2 + (1 - p)^2\right) \mathbf{E}(aL) + p(1 - p)\mathbf{E}(aL + 1 - a) + \\
&\quad p(1 - p)\mathbf{E}(-aL + 1 - a) \\
&= p^2 a\mathbf{E}L + (1 - p)^2 a\mathbf{E}L + p(1 - p)a\mathbf{E}L + p(1 - p)(1 - a) - \\
&\quad p(1 - p)a\mathbf{E}L + p(1 - p)(1 - a) \\
&= a\left(p^2 + (1 - p)^2\right)\mathbf{E}L + 2p(1 - p)(1 - a),
\end{aligned}
$$

so that

$$
\mathbf{E}L = \frac{2p(1 - p)(1 - a)}{1 - a\left(p^2 + (1 - p)^2\right)}.
$$

Now, when X_1 and X_2 are independently "uniform" on their common support, i.e., when $p = 1/2$, the previous equation becomes:

$$
\mathbf{E}L = \frac{1 - a}{2 - a},
$$

a decreasing function of a over $[0, 1/2]$. For $a = 1/2$, we are in the case of the first example and we have $\mathbf{E}L = 1/3$, which coincides with the value for the expectation of L given by formula (3.18), when k is set equal to one. For $a = 1/3$, we find that the expected length of a segment whose endpoints are drawn independently from a "uniform" distribution on the Cantor set is given by $\mathbf{E}L = 2/5$.

Finally, for $a = 0$, X_1 and X_2 are independent, discrete uniform random variables with point masses $1/2$ at 0 and 1, and $\mathbf{E}L = 1/2$. As is always the case, the invariance condition can also be applied to calculate, recursively, higher integer moments of L.

The unique, invariant distribution for the IFS with probabilities $\{\mathbb{R}; v_1(x) = (1/2)x, v_2(x) = (1/2)x + 1/2, v_3(x) = -(1/2)x + 1/2;\ q_1 = 1/2, q_2 = q_3 = 1/4\}$ of the first example is, as we have noticed, $\text{Beta}(1, 2)$, and the unique, invariant distribution for the IFS with probabilities $\{\mathbb{R}; w_1(x) = (1/2)x, w_2(x) = (1/2)x + 1/2; p_1 = p_2 = 1/2\}$, is uniform on $[0, 1]$, which can also be regarded as a $\text{Beta}(1, 1)$ distribution.

These facts raise the question of whether it is possible to construct other hyperbolic IFSs with probabilities, whose invariant distributions are related to the Beta family. In order to produce an answer, let us first recall that, if $X_1, \ldots, X_m$ is a random sample from a uniform distribution on $[0, 1]$, and $X_{(1)}, \ldots, X_{(m)}$ denote the order statistics of the sample, the distribution of $X_{(r)}$, $1 \le r \le m$, is $\text{Beta}(r, m - r + 1)$ (again, see David (1981) for details).

We shall now outline a method for constructing a hyperbolic IFS with probabilities on $\mathbb{R}^m$, whose unique invariant distribution coincides with the joint distribution of the order statistics of a sample of size m from a uniform distribution on $[0, 1]$. The corresponding marginal distributions will therefore be $\text{Beta}(r, m - r + 1)$, for $1 \le r \le m$.

To begin with, consider m independent Markov chains $\{X_n^1\}_{n=0}^{\infty}$, $\ldots, \{X_n^m\}_{n=0}^{\infty}$ associated with the IFS with probabilities $\{\mathbb{R}; w_1(x) = (1/2)x, w_2(x) = (1/2)x + 1/2; p_1 = p_2 = 1/2\}$. Each one of these chains admits, as its unique, invariant, limiting distribution, a uniform distribution on $[0, 1]$. Let

$$\left\{ X_n = \left(X_n^{(1)}, \ldots, X_n^{(m)} \right)^T \right\}_{n=0}^{\infty}$$

be the vector process obtained, at each step n, by rearranging $X_n^1, \ldots, X_n^m$ in increasing order. Since $\{X_n^1\}_{n=0}^{\infty}, \ldots, \{X_n^m\}_{n=0}^{\infty}$ are independent with uniform limiting distributions on $[0, 1]$, it follows that the limiting distribution of the vector process $\{X_n\}_{n=0}^{\infty}$ must coincide with the joint distribution of the order statistics of a sample of size m from a uniform distribution on $[0, 1]$.

We shall now see that the process $\{X_n\}_{n=0}^{\infty}$ can be regarded as the Markov chain associated with an IFS with probabilities on $\mathbb{R}^m$. The state X_n of the vector process at time n is determined by which transformations are applied to the various entries of the state X_{n-1} at time $n - 1$. More precisely, let $i_1 < \ldots < i_k$ be the indices of the entries of X_{n-1} to which w_1 is applied, and let $j_1 < \ldots < j_l$ be the indices of the entries of X_{n-1} to which w_2 is applied, where $k + l = m$. Then we have:

$$
X_n = \begin{pmatrix} X_n^{(1)} \\ \vdots \\ X_n^{(k)} \\ X_n^{(k+1)} \\ \vdots \\ X_n^{(k+l)} \end{pmatrix} = \begin{pmatrix} w_1\left(X_{n-1}^{(i_1)}\right) \\ \vdots \\ w_1\left(X_{n-1}^{(i_k)}\right) \\ w_2\left(X_{n-1}^{(j_1)}\right) \\ \vdots \\ w_2\left(X_{n-1}^{(j_l)}\right) \end{pmatrix} = v(X_{n-1}).
$$

This is so, since the relative ordering between any two points x and y in $[0, 1]$, with $x < y$, changes if and only if w_2 is applied to x and w_1 is applied to y.

It is easy to verify that $v(x)$ is an affine transformation on $\mathbb{R}^m$ of the form $v(x) = Ax + b$, where the vector b has the first k entries equal to 0 and the last l entries equal to $1/2$, while A is a matrix with entries $a_{1 i_1} = \ldots = a_{k i_k} = a_{(k+1) j_1} = \ldots = a_{(k+l) j_l} = 1/2$, and all other entries equal to zero. Since either w_1 or w_2 can be applied to each entry in X_{n-1}, independently and with probability $1/2$, there exist 2^m affine transformations v_i of this type, and each one of them can be applied independently and with probability $1/(2^m)$ to X_{n-1}. It follows that $\{X_n\}_{n=0}^{\infty}$ can be regarded as the Markov chain associated with the IFS with probabilities $\{\mathbb{R}^m; v_1, \ldots, v_{2^m}; q_1 = \ldots = q_{2^m} = 1/(2^m)\}$. The unique invariant distribution for the IFS must then coincide with the distribution of the order statistics of a sample of size m from a uniform distribution on $[0, 1]$, which has, as we noticed before, marginal distributions that belong to the Beta family.

It is clear from the above discussion that any Beta distribution with integer parameters can be generated as one of the marginal distributions of the unique, invariant distribution for an appropriate hyperbolic IFS with probabilities. For example, in the case when

$m = 2$, the four affine transformations in the IFS are given by:

$$v_1((x_1, x_2)^T) = \begin{pmatrix} 0.5 & 0 \\ 0 & 0.5 \end{pmatrix} \begin{pmatrix} x_1 \\ x_2 \end{pmatrix} + \begin{pmatrix} 0 \\ 0 \end{pmatrix},$$

$$v_2((x_1, x_2)^T) = \begin{pmatrix} 0.5 & 0 \\ 0 & 0.5 \end{pmatrix} \begin{pmatrix} x_1 \\ x_2 \end{pmatrix} + \begin{pmatrix} 0 \\ 0.5 \end{pmatrix},$$

$$v_3((x_1, x_2)^T) = \begin{pmatrix} 0 & 0.5 \\ 0.5 & 0 \end{pmatrix} \begin{pmatrix} x_1 \\ x_2 \end{pmatrix} + \begin{pmatrix} 0 \\ 0.5 \end{pmatrix},$$

and

$$v_4((x_1, x_2)^T) = \begin{pmatrix} 0.5 & 0 \\ 0 & 0.5 \end{pmatrix} \begin{pmatrix} x_1 \\ x_2 \end{pmatrix} + \begin{pmatrix} 0.5 \\ 0.5 \end{pmatrix},$$

and the two marginal distributions of the invariant distribution for the IFS with probabilities $\{\mathbb{R}^2; v_1, \ldots, v_4; q_1 = \ldots = q_4 = 1/4\}$ are Beta $(1, 2)$ and Beta $(2, 1)$.

Once again, the invariance condition can be used to calculate, recursively, the various moments of the invariant distribution. For instance, applying Equation (3.11), first with $k = 1$, $l = 0$, and then with $k = 0$, $l = 1$, we obtain the following system of linear equations in $\mathbf{E}X_1$ and $\mathbf{E}X_2$ (where $X = (X_1, X_2)^T$ is assumed to be distributed like the invariant distribution for the previous IFS):

$$\begin{cases} \mathbf{E}X_1 = \frac{1}{8}\mathbf{E}X_1 + \frac{1}{8}\mathbf{E}X_1 + \frac{1}{8}\mathbf{E}X_2 + \frac{1}{8}\mathbf{E}X_1 + \frac{1}{8}, \\ \mathbf{E}X_2 = \frac{1}{8}\mathbf{E}X_2 + \frac{1}{8}\mathbf{E}X_2 + \frac{1}{8} + \frac{1}{8}\mathbf{E}X_1 + \frac{1}{8} + \frac{1}{8}\mathbf{E}X_2 + \frac{1}{8}. \end{cases}$$

This system admits the unique solution $\mathbf{E}X_1 = 1/3$, $\mathbf{E}X_2 = 2/3$, and these values are, as expected, in agreement with the values of the first moments of a Beta $(1, 2)$ distribution and a Beta $(2, 1)$ distribution, respectively.

4

The Round-Off Process

In this chapter, we shall define a simple Markov chain, which will serve as a model of the computational approximations introduced by the implementation of the random iteration algorithm. In particular, we shall be concerned with the propagation of rounding errors associated with floating-point arithmetic. The proposed model will provide a description of the behavior of such errors that, some simplifying assumptions notwithstanding, appears to be fairly realistic.

An appealing feature of the approximating Markov chain, which henceforth will be referred to as the *round-off process*, lies in the fact that it gives rise to an iteration algorithm for image generation which is amenable to exact computer implementation. Another important characteristic of the round-off process is that images generated by its associated iterative algorithm are close, in a sense to be precisely specified in the sequel, to the corresponding images generated by the random iteration algorithm. This finding, which hinges on the fact that the contractivity of the IFS prevents rounding errors from accumulating beyond a certain point, is particularly valuable in practical applications, since it guarantees that an image generated by means of the random iteration algorithm provides an accurate rendering of the attractor of the associated IFS.

The main difference between the round-off process and the Markov chain originally associated with the IFS is the replacement of the continuous state space of the latter with an approximating discrete state space. For notational convenience, and bearing in mind the applications to computer graphics, we shall restrict our attention to IFSs in $\mathbb{R}^2$.

The remainder of the chapter is organized as follows. We shall first outline the discretization procedure and its connections with the modeling of rounding errors, define the round-off process, and describe its associated image generation algorithm. We shall then examine in detail the theoretical properties of the round-off version of a contractive transformation on the plane, the elementary building block for constructing the round-off process. Several examples of round-off versions of contractive transformations will be then analyzed and illustrated graphically. Finally, we shall examine the limiting behavior of the round-off process, as the discretization procedure is successively refined.

4.1.   Definition of the Round-Off Process

We begin our construction by laying over $\mathbb{R}^2$ a grid of square elements whose sides, parallel to the coordinate axes, have length $1/M$, with M a positive integer, and whose centers are placed at $(i/M, j/M)^T$, with i and j integers. Formally, we give the following definition.

Definition 4.1. *Let a positive integer M and integers i, j be given. The subset*

$$
P_M(i,j) \;=\; \left\{ (x_1, x_2)^T \in \mathbb{R}^2 \,\middle|\, \frac{i}{M} - \frac{1}{2M} \leq x_1 < \frac{i}{M} + \frac{1}{2M}, \right.
$$
$$
\left. \frac{j}{M} - \frac{1}{2M} \leq x_2 < \frac{j}{M} + \frac{1}{2M} \right\}
$$

of $\mathbb{R}^2$ is called the pixel *of center $(i/M, j/M)^T$ and size $1/M$.*

Remark 4.2. The collection $\mathcal{P}_M = \{P_M(i,j)\}$, with i and j integers, of all pixels forms a disjoint cover of $\mathbb{R}^2$, for any given positive integer M.

Remark 4.3. Although we have used the term *pixel* in Definition 4.1, we can think of each element $P_M(i,j)$ as an approximation to the floating-point representation of any real vector that belongs to $P_M(i,j)$ itself. More precisely, we would like to identify any real vector in $P_M(i,j)$ with the center $(i/M, j/M)^T$ of the pixel. The choice of the term *pixel* was motivated by the suggestive possibility, in practical applications, of determining M in such a way that the size of a grid element coincides with the actual pixel size for the employed graphics device.

The idea of approximating transformations taking on values in a Euclidean space by means of discretized versions taking on values in a grid constituting a partition of the original space has been considered in several scientific publications. Reasons for doing so range from the necessity of constructing simple models of the propagation of rounding errors in computer applications (see Lax (1971)) to the desire of providing a simplified analysis of the behavior of nonlinear dynamical systems (see Hsu (1987)).

The employed terminology, being usually related to the particular application being considered, is far from standard. For example, the grid elements, termed pixels in this book, are called cells in Hsu (1987), where the term cell transformation is then used to refer to what here will be called the round-off version of a given transformation.

At this point, a brief discussion of floating-point arithmetic is in order. The widely accepted Institute for Electrical and Electronic Engineers (IEEE) Standard for Binary Floating Point Arithmetic (see Kahan et al. (1979) and Stevenson (1981)) specifies that a single precision normalized number x be stored in 32 bits, using 1 bit to represent its sign s, 8 bits to represent its exponent e, and 23 bits to represent its fractional part f, in the following way. If $0 < e < 255$, then $x = (-1)^s \cdot 2^{e-127} \cdot (1.f)$; if $e = 0$ and $f \neq 0$, then $x = (-1)^s \cdot 2^{-126} \cdot (0.f)$; if $e = 0$ and $f = 0$, then $x = (-1)^s \cdot 0$. Notice that, when $0 < e < 255$, the digit to the left of the binary point is assumed to be 1 and is not stored. The exponent $e = 255$

is reserved for an object called Not-a-Number. As a consequence of these conventions, numbers close to zero in absolute value can be more precisely represented than numbers of larger absolute magnitude, thereby making our regular grid of pixels an approximate model only.

The IEEE standard also specifies that, when rounding to nearest, the result must differ from the infinite precision exact result by at most one half in the least-significant-digit position. It follows that, if the calculations involve only numbers that are less than one in absolute value, the worst possible rounding error resulting from an elementary floating-point operation is bounded, in absolute value, by 2^{-24}. Since the calculations to be performed at each step of the iterative algorithm that we are going to consider can often be regarded as elementary and, without loss of generality, as taking on values in $[-1, 1]$, a choice of $M = 2^{24}$ in Definition 4.1 will provide a pixel size suited for a worst-case analysis of the propagation of rounding errors, when single precision floating-point arithmetic is used.

Remark 4.3 suggests a way of modifying the random iteration algorithm to account for rounding errors. In order to do so, we need to restrict the codomain of a transformation $w : \mathbb{R}^2 \longmapsto \mathbb{R}^2$ to the set $\mathcal{K}_M = \{(i/M, j/M)^T\}$ of centers of pixels in $\mathcal{P}_M$. This is done, in the obvious way, as follows.

Definition 4.4. *Let $w : \mathbb{R}^2 \longmapsto \mathbb{R}^2$ and a positive integer M be given. The* round-off version *of w of accuracy $1/M$, $\tilde{w}^M : \mathbb{R}^2 \longmapsto \mathcal{K}_M$, is defined, for every $x \in \mathbb{R}^2$, as $\tilde{w}^M(x) = (i/M, j/M)^T$, where i and j are the unique integers such that $w(x) \in P_M(i, j)$.*

For notational simplicity, we shall often drop the superscript M in $\tilde{w}^M$, provided the pixel size can be easily inferred from the context.

Remark 4.5. If we regard $\mathcal{K}_M$ as a subset of $\mathbb{R}^2$, it follows immediately that the restriction of $\tilde{w}$ to $\mathcal{K}_M$ is well-defined. Since each element of $\mathcal{K}_M$ corresponds to one and only one element of $\mathcal{P}_M$, we can also think of $\tilde{w}$ as taking on values in $\mathcal{P}_M$, and regard its restriction to $\mathcal{K}_M$ as a transformation from $\mathcal{P}_M$ into $\mathcal{P}_M$.

We shall now make use of Definition 4.4 to provide a rigorous definition of the round-off process.

Definition 4.6. *Let* $\{\mathbb{R}^2; w_1, \ldots, w_N; p_1, \ldots, p_N\}$ *be a hyperbolic IFS with probabilities and let a positive integer* M *be given. The recursion*

$$\tilde{X}_{n+1} = \tilde{w}_{\sigma_n}(\tilde{X}_n), \qquad \text{for } n = 0, 1, 2, \ldots, \text{ and } \tilde{X}_0 \in \mathcal{K}_M,$$

where σ_n *is an integer random variable such that*

$$P\{\sigma_n = i\} = P\left\{\sigma_n = i \,\Big|\, \tilde{X}_0, \tilde{X}_1, \ldots, \tilde{X}_n\right\} = p_i, \quad \text{for } i = 1, \ldots, N,$$

and $\tilde{w}_{\sigma_n}$ *is the round-off version of* w_{σ_n} *of accuracy* $1/M$, *determines a Markov chain* $\{\tilde{X}_n\}_{n=0}^{\infty}$ *on* $\mathcal{K}_M$, *which is called the* round-off pro*cess of accuracy* $1/M$ *associated with the IFS* $\{\mathbb{R}^2; w_1, \ldots, w_N; p_1, \ldots, p_N\}$. *Whenever the dependence of the round-off process on the pixel size needs to be made explicit, we shall employ the notation* $\{\tilde{X}_n^M\}_{n=0}^{\infty}$.

Remark 4.7. We can think of $\{\tilde{X}_n\}_{n=0}^{\infty}$ as of the Markov chain associated with the *discrete* IFS $\{\mathcal{K}_M; \tilde{w}_1, \ldots, \tilde{w}_N; p_1, \ldots, p_N\}$. Notice that, even if the original IFS is hyperbolic, the corresponding discrete IFS need not be (see comment on page 80).

Based on this chain, we have the following discretized analogue of the random iteration algorithm for image generation. For any initial pixel $\tilde{x}_0 \in \mathcal{K}_M$, randomly select, independently of $\tilde{x}_0$, $\sigma_0 \in \{1, \ldots, N\}$, according to the probabilities $(p_1, \ldots, p_N)$, and compute $\tilde{x}_1 = \tilde{w}_{\sigma_0}(\tilde{x}_0)$. Iterate the procedure, so that, at stage n, $\tilde{x}_{n+1} = \tilde{w}_{\sigma_n}(\tilde{x}_n)$ is computed, where $\sigma_n \in \{1, \ldots, N\}$ has been chosen at random according to the given probabilities and independently of all the points that have been previously generated.

The sequence $\{\tilde{x}_n\}_{n=0}^{\infty}$ constitutes a trajectory of the Markov chain $\{\tilde{X}_n\}_{n=0}^{\infty}$, and could, in principle, be used to generate a digitized image, in much the same way in which we use the trajectory $\{x_n\}_{n=0}^{\infty}$ of the Markov chain $\{X_n\}_{n=0}^{\infty}$, when applying the original random iteration algorithm.

We recall, here, that when the IFS $\{\mathbb{R}^2; w_1, \ldots, w_N\}$ is hyperbolic, the Markov chain $\{X_n\}_{n=0}^{\infty}$ is ergodic, in the sense that, for any starting point $x_0 \in \mathbb{R}^2$, the empirical distributions of almost every trajectory $\{x_n\}_{n=0}^{\infty}$ converge weakly to the unique stationary probability ν of the chain. Ergodicity of the chain is actually key to a successful implementation of the algorithm, since it guarantees that, for a given set of mappings and associated probabilities, the same image will always be obtained, regardless of the starting point and of the particular sequence of mappings being selected.

At this point, there are a number of questions that arise naturally and need to be answered. We have mentioned before that we could employ orbits of the round-off process (that can actually be calculated exactly by a computer) for the purpose of image generation. In order to do so, though, we would like to be able to assert that such orbits behave in a manner similar to that of their continuous counterparts. In particular, it would be desirable that different orbits of $\{\tilde{X}_n\}_{n=0}^{\infty}$ generated the same image. Unfortunately, this is not always the case, since there exist cases in which the round-off process does not possess a unique stationary distribution.

The situation, however, is not hopeless. In fact, once the accuracy $1/M$ has been fixed, the collection of recurrent states of the process $\{\tilde{X}_n^M\}_{n=0}^{\infty}$ turns out to be finite. This in turn implies that the number of classes of communicating states is itself finite and each such class has a unique stationary distribution. Therefore, for a given initial condition $\tilde{x}_0 \in \mathcal{K}_M$, the empirical distributions of almost every orbit converge to one of these stationary distributions, e.g. $\pi_{k(M)}$. It is then possible to show that any sequence of distributions $\{\pi_{k(M)}\}_{M=1}^{\infty}$, constructed by choosing, at each stage M, one among the finitely many stationary distributions for $\{\tilde{X}_n^M\}_{n=0}^{\infty}$, converges weakly, as M tends to infinity, to ν, the unique stationary distribution of the original process $\{X_n\}_{n=0}^{\infty}$. At the same time, the sequence of their respective supports converges to the support A of ν in the Hausdorff metric.

In terms of our applications to computer graphics, these results tell us that, for any given value of M, there are only a finite number of different images that can be generated by the random iteration algorithm based on the round-off process. Furthermore, all these images are approximations to the image that would be obtained by applying the random iteration algorithm based on the true chain $\{X_n\}_{n=0}^{\infty}$, and they are good approximations, in the sense that they tend to get closer to the true image as the value of M increases.

4.2. Properties of the Round-Off Version of a Single Contractive Transformation on $\mathbb{R}^2$

In order to prove the results outlined at the end of the previous section, we need to understand the behavior of the round-off version of a contractive transformation of $\mathbb{R}^2$ into itself. In particular, we are interested in studying orbits generated by starting at any point in $\mathcal{K}_M$ and repeatedly applying the round-off version $\tilde{w}$ of a contraction w. As noted in Remark 4.5, we can equivalently regard these orbits as taking on values in $\mathcal{K}_M$ or $\mathcal{P}_M$.

In order to describe such orbits, we shall need to make use of a metric on $\mathcal{K}_M$. Since $\mathcal{K}_M$ is a subset of $\mathbb{R}^2$, it can be made into a metric space by defining the distance between any two of its points to be the distance between those two points when regarded as elements of the metric space $(\mathbb{R}^2, d)$, where d is any given metric on $\mathbb{R}^2$. Unless otherwise stated, we shall assume that the metric d on $\mathbb{R}^2$ coincides with the Euclidean metric. It is also an immediate consequence of Remark 4.5 that the spaces $(\mathcal{K}_M, d)$ and $(\mathcal{P}_M, d)$ are isometric and therefore interchangeable for the purpose of our discussion. This justifies our exclusive use of the symbol $\mathcal{P}_M$ (or $\mathcal{P}$ when the pixel size can be easily inferred from the context) from this point on.

Throughout the remainder of this section, we shall make use of the following assumptions:

(A1) $w : \mathbb{R}^2 \longmapsto \mathbb{R}^2$ is a strict contraction, with contractivity factor s.

(A2) The fixed point of w coincides with the origin $O = (0,0)^T$ of the coordinate system.

Note that the condition of the second assumption can always be met, without any substantial loss of generality, by applying an appropriate translation.

We recall that successive iterates of a contraction w, starting at any initial point $x_0 \in \mathbb{R}^2$, converge to the fixed point of w. The following lemma will be used in the sequel to prove that an analogous result holds for $\tilde{w}$, the round-off version of w.

Lemma 4.8. *Assume that (A1) and (A2) hold. Let $\tilde{x}_0$ be any element of $\mathcal{P}_M$ and define recursively:*

$$\tilde{x}_{n+1} = \tilde{w}(\tilde{x}_n) = \tilde{w}^{\circ(n+1)}(\tilde{x}_0), \qquad\qquad \text{for } n = 0, 1, 2, \ldots,$$

where $\tilde{w}$ is the round-off version of w of accuracy $1/M$ (here and in what follows, $\tilde{w}^{\circ n}$ denotes the n-fold composition of $\tilde{w}$ with itself). Let $D = d(\tilde{x}_0, O)$ and $\theta = 1/(M\sqrt{2})$. Then

$$d\left(w(\tilde{x}_n), O\right) \leq \theta s \frac{1 - s^n}{1 - s} + s^{n+1} D, \qquad \text{for } n = 0, 1, 2, \ldots. \qquad (4.1)$$

Proof. We shall prove the result by induction. Since w is a contraction with contractivity factor s and its fixed point coincides with the origin O, we have:

$$d\left(w(\tilde{x}_0), O\right) \leq s\, d(\tilde{x}_0, O) = sD,$$

so that inequality (4.1) is true for $n = 0$. Assume inequality (4.1) is true for n. Using again the fact that w is a contraction whose fixed point coincides with the origin, and applying the triangle inequality, we have:

$$d\left(w(\tilde{x}_{n+1}), O\right) \leq s\, d\left(\tilde{x}_{n+1}, O\right) \leq s\left[d\left(\tilde{x}_{n+1}, w(\tilde{x}_n)\right) + d\left(w(\tilde{x}_n), O\right)\right].$$

Since $\tilde{x}_{n+1}$ is defined to be the center of the pixel to which $w(\tilde{x}_n)$ belongs, $d\left(\tilde{x}_{n+1}, w(\tilde{x}_n)\right)$ can be bounded by half the length of the pixel diagonal. Applying this fact and the induction hypothesis to the last term in the previous chain of inequalities, we have:

$$
\begin{aligned}
d\left(w(\tilde{x}_{n+1}), O\right) &\leq s\left[\theta + \theta s \frac{1 - s^n}{1 - s} + s^{n+1} D\right] \\
&= \theta s\left[\frac{1 - s + s - s^{n+1}}{1 - s}\right] + s^{n+2} D \\
&= \theta s \frac{1 - s^{n+1}}{1 - s} + s^{n+2} D,
\end{aligned}
$$

so that inequality (4.1) is also true for $n + 1$, which completes the proof.

Our next goal is to show that there exists a subset of $\mathcal{P}_M$ that acts like an attractive fixed point for the round-off version of a given contraction w, in the sense of being the smallest subset of $\mathcal{P}_M$ that will eventually contain every orbit under $\tilde{w}$. The following definition will help make these notions more precise.

Definition 4.9. *Let $w : \mathbb{R}^2 \longmapsto \mathbb{R}^2$ be a given transformation, and let $\tilde{w}$ be its round-off version of accuracy $1/M$. We say that a set $A \subset \mathcal{P}_M$ is absorbing for $\tilde{w}$ if and only if, for any starting pixel $\tilde{x}_0 \in \mathcal{P}_M$, there exists a nonnegative integer $n_0 = n_0(\tilde{x}_0)$ such that $\tilde{x}_n = \tilde{w}^{\circ n}(\tilde{x}_0) \in A$, for every $n \geq n_0$.*

Remark 4.10. Given any transformation w, the set $\mathcal{P}_M$ of all pixels is obviously absorbing for $\tilde{w}$, since Definition 4.9 is satisfied if we take $n_0 = 0$ for any starting point.

In our context, interesting transformations are those whose round-off versions possess absorbing sets other than the set of all pixels. Strictly contractive transformations of $\mathbb{R}^2$ into itself have this property, as the following lemma illustrates.

Lemma 4.11. *Assume that (A1) and (A2) hold, and let $\tilde{w}$ be the round-off version of w of accuracy $1/M$. Then, there exists a set $A \subset \mathcal{P}_M$ which is absorbing for $\tilde{w}$ and has finite cardinality.*

Proof. The right-hand side of inequality (4.1) converges to $(\theta s)/(1 - s)$, as n goes to infinity. Therefore, the distance between $w(\tilde{x}_n)$ and the closure of $B(O, (\theta s)/(1 - s))$, the open ball of center O and radius $(\theta s)/(1 - s)$, will be arbitrarily small for n large enough. It follows that, for any starting pixel $\tilde{x}_0$, there exists a nonnegative integer $n_0 = n_0(\tilde{x}_0)$ such that, for any $n \geq n_0$, $\tilde{x}_n$ must coincide with one of the pixels whose closure (when the pixels are viewed as subsets of $\mathbb{R}^2$ endowed with the Euclidean metric) has nonempty intersection with $\overline{B(O, (\theta s)/(1 - s))}$. The set A of pixels that have such a property is therefore absorbing and obviously finite, which completes the proof.

The following corollary provides a bound on the cardinality of $\mathcal{A}$ that depends on the contractivity factor s of w, but is independent of the level of accuracy $1/M$.

Corollary 4.12. *Under the same assumptions as in Lemma 4.11, the cardinality of $\mathcal{A}$ is less than, or equal to,*

$$K(s) = \left(2 \left\lfloor \frac{s}{(1-s)\sqrt{2}} + 1/2 \right\rfloor + 1 \right)^2,$$

where the symbol $\lfloor \cdot \rfloor$ denotes the integer part of its argument.

Proof. Observe that $\overline{B\left(O, (\theta s)/(1-s)\right)}$ is contained in the square

$$S = \left\{ (x,y)^T \in \mathbb{R}^2 \,\middle|\, 0 \le |x|, |y| \le (\theta s)/(1-s) \right\},$$

so that, if the closure of a pixel has nonempty intersection with $\overline{B\left(O, (\theta s)/(1-s)\right)}$, it must also have nonempty intersection with S. Hence, the cardinality of $\mathcal{A}$ is no greater than the cardinality of the set of pixels whose closure has nonempty intersection with S. In order for a pixel $P(i,j)$ to have nonempty intersection with S, we must have:

$$-\frac{1}{2M} \le \frac{|i|}{M} - \frac{1}{2M} \le \frac{\theta s}{1-s} = \frac{1}{M} \frac{s}{(1-s)\sqrt{2}},$$

and

$$-\frac{1}{2M} \le \frac{|j|}{M} - \frac{1}{2M} \le \frac{\theta s}{1-s} = \frac{1}{M} \frac{s}{(1-s)\sqrt{2}}.$$

These two inequalities are equivalent to:

$$0 \le |i|, |j| \le \left\lfloor \frac{s}{(1-s)\sqrt{2}} + \frac{1}{2} \right\rfloor.$$

Since there are

$$\left(2 \left\lfloor \frac{s}{(1-s)\sqrt{2}} + 1/2 \right\rfloor + 1 \right)^2$$

pairs of integers i, j that satisfy this last condition, the result follows.

The following lemma addresses the question of what happens when we intersect all possible absorbing sets for the round-off version of a contractive transformation of the Euclidean plane into itself, and provides the justification for the introduction of a notion of minimality for absorbing sets.

Lemma 4.13. *Assume that (A1) and (A2) hold, and let $\{\mathcal{A}_\alpha\}_{\alpha \in I}$ be the collection of all absorbing sets for $\tilde{w}$, the round-off version of w of accuracy $1/M$. Then, $\mathcal{M} = \bigcap_{\alpha \in I} \mathcal{A}_\alpha$ is a nonempty subset of $\mathcal{P}_M$, has finite cardinality, and is absorbing for $\tilde{w}$.*

Proof. We first note that $\mathcal{M}$ is nonempty, since pixel $O = P(0, 0)$ belongs to $\mathcal{A}_\alpha$, for any $\alpha \in I$. This follows directly from the fact that $\tilde{w}^{on}(O) = O$, for every nonnegative integer n, and from the definition of an absorbing set.

Lemma 4.11 guarantees the existence of an absorbing set $\mathcal{A}_{\bar{\alpha}}$ of finite cardinality. Since $\mathcal{A}_{\bar{\alpha}}$ contains $\mathcal{M}$, it follows that the cardinality of $\mathcal{M}$ is finite.

It remains to be shown that $\mathcal{M}$ is absorbing for $\tilde{w}$. We first show that any finite intersection of absorbing sets is an absorbing set. Suppose, in fact, that $\mathcal{A}_1, \mathcal{A}_2, \ldots, \mathcal{A}_m$ are absorbing for $\tilde{w}$. Then for any starting point $\tilde{x}_0 \in \mathcal{P}_M$, there exists a nonnegative integer n_{0i} such that, for every $n \geq n_{0i}$, $\tilde{x}_n = \tilde{w}^{on}(\tilde{x}_0) \in \mathcal{A}_i$, for $1 \leq i \leq m$. Hence, for every $n \geq n_0 = \max_{1 \leq i \leq m} n_{0i}$, we have $\tilde{x}_n = \tilde{w}^{on}(\tilde{x}_0) \in \bigcap_{i=1}^m \mathcal{A}_i$, so that $\bigcap_{i=1}^m \mathcal{A}_i$ is absorbing for $\tilde{w}$.

The proof will be completed if we can show that $\mathcal{M} = \bigcap_{\alpha \in I} \mathcal{A}_\alpha$ can be expressed as a finite intersection of absorbing sets. Let $\mathcal{A}_{\bar{\alpha}}$ be the absorbing set of finite cardinality introduced in the first part of the proof, and define $\mathcal{B}_\alpha = \mathcal{A}_\alpha \cap \mathcal{A}_{\bar{\alpha}}$, for any $\alpha \in I$. Then, for any $\alpha \in I$, $\mathcal{B}_\alpha$ is absorbing for $\tilde{w}$, being the intersection of two absorbing sets. Moreover, $\mathcal{M} = \bigcap_{\alpha \in I} \mathcal{A}_\alpha = \bigcap_{\alpha \in I} \mathcal{B}_\alpha$. Now, each $\mathcal{B}_\alpha$ is a subset of $\mathcal{A}_{\bar{\alpha}}$. Since $\mathcal{A}_{\bar{\alpha}}$ has finite cardinality, it can only have a finite number of different subsets. It follows that there exist only finitely many sets $\mathcal{B}_\alpha$ which are distinct. Hence, $\mathcal{M}$ can be regarded as a finite intersection of absorbing sets and is therefore absorbing for $\tilde{w}$.

Definition 4.14. *The subset $\mathcal{M}$ of $\mathcal{P}_M$ defined in Lemma 4.13 is called* minimal *absorbing for $\tilde{w}$.*

Remark 4.15. It follows from Corollary 4.12 that the cardinality of $\mathcal{M}$ can be bounded above, independently of the level of accuracy $1/M$, by an integer $K(s)$ which depends only on the contractivity factor s of w.

A few comments are in order. When assumptions (A1) and (A2) are satisfied, so that w is a strict contraction whose unique fixed point coincides with the origin $O = (0,0)^T$ of the coordinate system, the pixel $P(0,0)$ centered at the origin is a fixed point of $\tilde{w}$. The fact that w is contractive, however, does not necessarily imply that $\tilde{w}$ is also contractive. Indeed, $\tilde{w}$ might have other fixed points besides O, and O need not be attractive for $\tilde{w}$. This constitutes a fundamental difference between $\tilde{w}$ and w, which has instead a unique fixed point to which every orbit converges.

On the other hand, $\tilde{w}$ being so intimately related to w, we would expect its orbits to show a behavior that still resembles that of orbits under w. This is actually the case when we regard $\mathcal{M}$, the minimal absorbing set for $\tilde{w}$, as the discrete counterpart to the fixed point of w. For one thing, every orbit under $\tilde{w}$, regardless of its starting condition, will eventually belong to $\mathcal{M}$, which is a nonempty, finite subset of the range of all possible values for orbits under $\tilde{w}$. Furthermore, $\mathcal{M}$ is the smallest subset of $\mathcal{P}_M$ that will eventually absorb all orbits under $\tilde{w}$.

This parallel can be carried even further, as we shall show that $\mathcal{M}$ is a fixed set for $\tilde{w}$, in the sense that $\tilde{w}(\mathcal{M}) = \mathcal{M}$ and $\tilde{w}^{-1}(\mathcal{M}) = \mathcal{M}$. In order to do so, we give the following definition and prove a lemma that provides a different characterization of $\mathcal{M}$.

Definition 4.16. *Let $w : \mathbb{R}^2 \longmapsto \mathbb{R}^2$ be given, and let $\tilde{w}$ be its round-off version of accuracy $1/M$. We say that the orbit of a point $\tilde{x}_0 \in \mathcal{P}_M$ under $\tilde{w}$ visits a subset $\mathcal{A}$ of $\mathcal{P}_M$ infinitely often (i.o.) if and only if, for any nonnegative integer n_0, there exists an integer $n \geq n_0$ such that $\tilde{w}^{\circ n}(\tilde{x}_0) \in \mathcal{A}$.*

Lemma 4.17. *Suppose that (A1) and (A2) hold, and let $\tilde{w}$ be the*

Color Plate 1

Color Plate 2

Color Plate 3

Color Plate 4

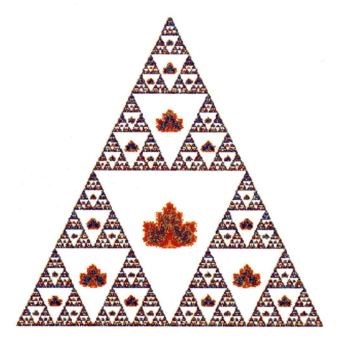

Color Plate 5

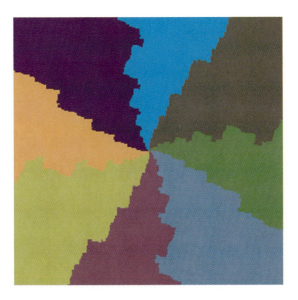

Color Plate 6

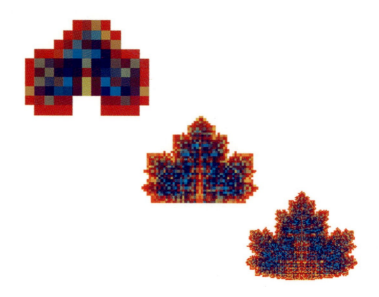

Color Plate 7

Color Plate 8

round-off version of w of accuracy $1/M$. A necessary and sufficient condition for a point $\tilde{x} \in \mathcal{P}_M$ to belong to the minimal absorbing set $\mathcal{M}$ is that there exists a point $\tilde{x}_0 \in \mathcal{P}_M$ such that the orbit of $\tilde{x}_0$ under $\tilde{w}$ will visit $\{\tilde{x}\}$ infinitely often.

Proof. We shall prove that the condition is necessary by contradiction. Suppose that $\tilde{x} \in \mathcal{M}$ and that every orbit under $\tilde{w}$ will not visit $\{\tilde{x}\}$ i.o. Then, for any $\tilde{x}_0 \in \mathcal{P}_M$, there exists a nonnegative integer n_0 such that, for any $n \geq n_0$, $\tilde{w}^{on}(\tilde{x}_0) \neq \tilde{x}$. Since $\mathcal{M}$ is absorbing, there exists a nonnegative integer m_0 such that, for any $n \geq m_0$, $\tilde{w}^{on}(\tilde{x}_0) \in \mathcal{M}$. Hence, for any $n \geq \max(n_0, m_0)$, $\tilde{w}^{on}(\tilde{x}_0) \in \mathcal{M} \setminus \{x\}$, so that $\mathcal{M} \setminus \{x\}$ is absorbing for $\tilde{w}$. This, however, contradicts the minimality property of $\mathcal{M}$.

Conversely, suppose there exists a point $\tilde{x}_0 \in \mathcal{P}_M$ such that the orbit of $\tilde{x}_0$ under $\tilde{w}$ will visit $\{\tilde{x}\}$ i.o., and let $\mathcal{A}$ be any absorbing set for $\tilde{w}$. If $\tilde{x}$ did not belong to $\mathcal{A}$, the orbit in question could not eventually stay in $\mathcal{A}$, and $\mathcal{A}$ would not be absorbing. Thus, $\tilde{x}$ must belong to $\mathcal{A}$ and, since $\mathcal{A}$ is an arbitrary absorbing set for $\tilde{w}$, $\tilde{x}$ must also belong to $\mathcal{M}$.

The following lemma provides a formal statement of the assertion, previously made, that $\mathcal{M}$ can be regarded as a fixed set of $\tilde{w}$.

Lemma 4.18. *Assume that (A1) and (A2) are satisfied, and let $\tilde{w}$ be the round-off version of w of accuracy $1/M$. The restriction of $\tilde{w}$ to the minimal absorbing set $\mathcal{M}$ is a bijection of $\mathcal{M}$ into itself.*

Proof. We first need to show that $\tilde{w}$ maps $\mathcal{M}$ into itself. Suppose that $\tilde{x} \in \mathcal{M}$ is given and let $\tilde{y} = \tilde{w}(\tilde{x})$. Since the condition in Lemma 4.17 is necessary, there exists an orbit under $\tilde{w}$ that visits $\{\tilde{x}\}$ i.o. Obviously, such an orbit must also visit $\tilde{y}$ i.o. The sufficiency of the condition in Lemma 4.17 implies then that $\tilde{y}$ belongs to $\mathcal{M}$.

The restriction of $\tilde{w}$ to $\mathcal{M}$ is surjective. Let $\tilde{y} \in \mathcal{M}$ be given. Once again, Lemma 4.17 implies that there exists an $\tilde{x}_0 \in \mathcal{P}_M$ such that the orbit of $\tilde{x}_0$ under $\tilde{w}$ will visit $\{\tilde{y}\}$ i.o. Since $\mathcal{M}$ is absorbing, there exists a nonnegative integer n_0 such that, for any $n \geq n_0$, $\tilde{w}^{on}(\tilde{x}_0) \in \mathcal{M}$. Then, if $\tilde{w}^{-1}(\{\tilde{y}\}) \bigcap \mathcal{M} = \{\tilde{x} \in \mathcal{M} | \tilde{w}(\tilde{x}) = \tilde{y}\}$ were empty, we would have $\tilde{w}^{\circ n}(\tilde{x}_0) \neq \tilde{y}$ for any $n > n_0$, and the orbit of $\tilde{x}_0$ under $\tilde{w}$ could not visit $\tilde{y}$ i.o. Hence, there must exist at least one element $\tilde{x} \in \mathcal{M}$ such that $\tilde{w}(\tilde{x}) = \tilde{y}$.

Finally, the restriction of $\tilde{w}$ to $\mathcal{M}$ is injective, since any surjective transformation of a set of finite cardinality into itself must be so.

Like any other absorbing set, $\mathcal{M}$ will eventually contain every orbit under $\tilde{w}$. Its minimality property, however, makes it special in many respects. It is an immediate consequence of the previous lemma, for instance, that once an orbit under $\tilde{w}$ has entered $\mathcal{M}$, it will never leave it, while Lemma 4.17 guarantees that, for any pixel in $\mathcal{M}$, there exists at least one orbit under $\tilde{w}$ that will go back to that pixel over and over. The following lemma implies that every orbit that visits a pixel $\tilde{x} \in \mathcal{M}$ will return to the same pixel infinitely often at regular intervals of time.

Lemma 4.19. *Assume that (A1) and (A2) are satisfied. Let $\tilde{w}$ denote the round-off version of w of accuracy $1/M$, and let m be the cardinality of $\mathcal{M}$. Then, for any $\tilde{x} \in \mathcal{M}$, there exists an integer n, with $1 \leq n \leq m$, such that $\tilde{w}^{\circ n}(\tilde{x}) = \tilde{x}$.*

Proof. Let $\tilde{x} \in \mathcal{M}$ be given. It follows from Lemma 4.17 that there exists an orbit of some pixel $\tilde{x}_0$ under $\tilde{w}$ that visits $\{\tilde{x}\}$ i.o. Let n_1 and n_2 denote the first and the second time that such an orbit visits $\{\tilde{x}\}$, and let $n = n_2 - n_1$. Then we have:

$$\tilde{w}^{\circ n}(\tilde{x}) = \tilde{w}^{\circ(n_2-n_1)}(\tilde{x}) = \tilde{w}^{\circ(n_2-n_1)}(\tilde{w}^{\circ n_1}(\tilde{x}_0)) = \tilde{w}^{\circ n_2}(\tilde{x}_0) = \tilde{x}.$$

We must have $n \leq m$, or else, if we had $n > m$, the following contradictory situation would arise. Let $\tilde{y}_i = \tilde{w}^{\circ i}(\tilde{x})$, for $1 \leq i \leq n-1$ (observe that $n-1 \geq 1$, since $m \geq 1$). Lemma 4.18 implies that every $\tilde{y}_i$ belongs to $\mathcal{M}$. Furthermore, since the orbit of $\tilde{x}_0$ under $\tilde{w}$ does not visit $\{\tilde{x}\}$ at any time between n_1 and n_2, we must have:

$$\tilde{x} \neq \tilde{w}^{\circ(n_1+i)}(\tilde{x}_0) = \tilde{w}^{\circ i}(\tilde{x}) = \tilde{y}_i, \qquad \text{for } 1 \leq i \leq n - 1.$$

We now distinguish two cases. If $m = 1$, the previous relation cannot hold, since $\mathcal{M}$ contains only one element. Consider, then, the case when $m > 1$, so that $n - 1 \geq 2$. The assumption that the cardinality of $\mathcal{M}$ is strictly smaller than n implies that $\mathcal{M}$ contains at most $n-1$ distinct pixels. Therefore, since every $\tilde{y}_i$ belongs to $\mathcal{M}$ and differs from $\tilde{x}$, which is also in $\mathcal{M}$, at least one of the $\tilde{y}_i$ must be repeated, as i ranges between 1 and $n-1$. In other words, there exist two integers j and k, with $1 \leq j < k \leq n-1$, such that $\tilde{y}_j = \tilde{y}_k$. Then we have:

$$\tilde{w}^{\circ(k-j)}(\tilde{y}_j) = \tilde{w}^{\circ(k-j)}(\tilde{w}^{\circ j}(\tilde{x})) = \tilde{w}^{\circ k}(\tilde{x}) = \tilde{y}_k = \tilde{y}_j.$$

This, together with the fact that $\tilde{y}_i \neq \tilde{x}$, for $1 \leq i \leq k$, makes it impossible for the orbit of $\tilde{x}_0$ under $\tilde{w}$ ever to return to $\{\tilde{x}\}$ after n_1.

Remark 4.20. The value of n determined in the proof of the previous lemma is the smallest integer for which $\tilde{w}^{\circ n}(\tilde{x}) = \tilde{x}$.

Lemma 4.19 and the previous remark warrant the introduction of the two definitions that follow.

Definition 4.21. *Suppose that (A1) and (A2) are satisfied. Let $\tilde{w}$ denote the round-off version of w of accuracy $1/M$, and let $\tilde{x}$ be a pixel belonging to the minimal absorbing set $\mathcal{M}$. The smallest integer n for which $\tilde{w}^{\circ n}(\tilde{x}) = \tilde{x}$ is called the* period *of $\tilde{x}$ under $\tilde{w}$.*

Definition 4.22. *Under the same assumptions as in Definition 4.21, suppose further that a pixel $\tilde{x}$ belonging to $\mathcal{M}$ has period n. The set*

$$\mathcal{C}_{\tilde{x}} = \left\{ \tilde{y}_i \in \mathcal{M} | \tilde{y}_i = \tilde{w}^{\circ i}(\tilde{x}), i = 0, 1, \ldots, n-1 \right\}$$

is called the (minimal absorbing) component *for $\tilde{w}$ generated by $\tilde{x}$.*

Remark 4.23. It is clear that every component can be regarded as being generated by any of its elements. Different components are disjoint and their union coincides with the minimal absorbing set $\mathcal{M}$. In other words, the set $\mathcal{M}$ can be uniquely partitioned into a finite number of disjoint, nonempty components. Furthermore, all pixels in a given component have a common period.

Remark 4.24. Since the number of minimal absorbing components for $\tilde{w}$ can be no greater than the cardinality of the minimal absorbing set $\mathcal{M}$, it follows immediately from Remark 4.15 that such a number can be bounded above by the integer $K(s)$ defined in Corollary 4.12. The said integer is independent of the level of accuracy $1/M$, and is only a function of the contractivity factor s of w.

The actual cardinality of the minimal absorbing set $\mathcal{M}$ and the number of components into which it can be partitioned are obviously dependent on the contraction w, and on the level of accuracy $1/M$. The following lemma relates the cardinality of $\mathcal{M}$ to the number of its components in one important, special case.

Lemma 4.25.   *Assume that (A1) and (A2) hold. Let $\mathcal{M}$ be the minimal absorbing set for $\tilde{w}$, the round-off version of w of accuracy $1/M$. Then, the cardinality of $\mathcal{M}$ equals one if and only if there exists only one minimal absorbing component. Such a component is the one consisting of pixel $O = P(0,0)$.*

Proof.   As we noted in the proof of Lemma 4.13, pixel $O = P(0,0)$ always belongs to $\mathcal{M}$. Since $\tilde{w}(O) = O$, O has period one and, therefore, $\mathcal{C}_O = \{O\}$. Hence, if the cardinality of $\mathcal{M}$ is one, $\mathcal{C}_O$ must be the only component. Conversely, if there exists only one component, it necessarily coincides with $\mathcal{C}_O$, so that the cardinality of $\mathcal{M}$ is one.

One interesting consequence of the previous lemma is that, whenever $\mathcal{M}$ contains more than one pixel, it can be partitioned into at least two components.

The case when $\mathcal{M}$ contains only one pixel will turn out to be of particular interest, as we shall see that, if an IFS contains a transformation whose round-off version possesses a minimal absorbing set of cardinality one, the round-off process associated with the given IFS has a unique stationary distribution. The following lemma provides a sufficient condition for the cardinality of $\mathcal{M}$ to be one.

Lemma 4.26.   *Let $w : \mathbb{R}^2 \longmapsto \mathbb{R}^2$ be a contraction, with contractivity factor s, $0 \le s < 1/2$. Suppose that (A2) is satisfied, and let $\tilde{w}$ be the round-off version of w of accuracy $1/M$. Then, the cardinality of $\mathcal{M}$, the minimal absorbing set for $\tilde{w}$, equals one.*

Proof.   It is enough to show that the set $\{O = P(0,0)\}$ is absorbing, since this will imply that $\mathcal{M}$, being the intersection of all absorbing sets, coincides with $\{O\}$. Let $\tilde{x}_0 \in \mathcal{P}_M$ be given. We need to bound the distance of $\tilde{x}_n = \tilde{w}^{on}(\tilde{x}_0)$ from the origin. Applying Equation (4.1) of Lemma 4.8 and the fact that $0 \leq 1 - s^n \leq 1$, we have:

$$d\left(w(\tilde{x}_n), O\right) \leq \theta \frac{s}{1-s} + s^{n+1}D, \qquad \text{for } n = 0, 1, 2, \ldots,$$

where $\theta = 1/(M\sqrt{2})$ and $D = d(\tilde{x}_0, O)$. Since $0 \leq s < 1/2$, $\theta s/(1-s)$ is strictly less than θ. Furthermore, $s^{n+1}D$ decreases monotonically to zero, as n goes to infinity. It follows that there exists a nonnegative integer n_0 such that $d(w(\tilde{x}_n), O) < \theta$, for any $n \geq n_0$. It is clear (see Figure 4.1) that, since $d(w(\tilde{x}_{n_0}), O) < \theta$, $\tilde{x}_{n_0+1}$ must be one of the pixels which have one edge in common with O, or O itself. This implies that $d(\tilde{x}_{n_0+1}, O) \leq 1/M$, so that $d(w(\tilde{x}_{n_0+1}), O) \leq sd(\tilde{x}_{n_0+1}, O) < 1/(2M)$. Thus, $w(\tilde{x}_{n_0+1})$ is contained in the pixel centered at the origin and $\tilde{x}_{n_0+2}$ coincides with O. It follows immediately that $\tilde{x}_n = \tilde{w}^{on}(\tilde{x}_0)$ coincides with O, for any $n \geq n_0 + 2$, so that $\{O\}$ is absorbing for $\tilde{w}$.

In light of the theory that has so far been developed, the typical behavior of an orbit under the round-off version of a contractive transformation of the Euclidean plane into itself can be summarized as follows. Starting at any pixel $\tilde{x}_0 \in \mathcal{P}_M$, the sequence $\{\tilde{x}_n = \tilde{w}^{on}(\tilde{x}_0)\}_{n=0}^{\infty}$ visits various pixels in $\mathcal{P}_M$ until it first enters the minimal absorbing set $\mathcal{M}$. From that point on, the orbit will keep visiting pixels in $\mathcal{M}$, without ever leaving it. Furthermore, the pixels in $\mathcal{M}$ that are visited by the orbit constitute one of the components into which $\mathcal{M}$ can be partitioned, and the orbit circulates between them at regular intervals of time.

The partition of $\mathcal{M}$ into a finite number of minimal absorbing components induces a partition of the space of pixels $\mathcal{P}_M$ into an equal number of nonempty regions, characterized by the property that all orbits starting in one region will eventually be absorbed by the same component. We make this notion more precise by giving the following definition.

Definition 4.27. *Assume that (A1) and (A2) hold. Let $\tilde{w}$ be the*

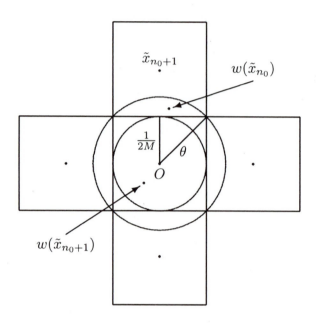

Figure 4.1. Converging to O: If w has a contractivity factor smaller than $1/2$, every orbit under $\tilde{w}$ will eventually end up in one of the five pixels depicted above. It is then enough to apply $\tilde{w}$ one more time to be certain that the orbit is absorbed by O.

round-off version of w of accuracy $1/M$, and let $\mathcal{M}$ be the minimal absorbing set for $\tilde{w}$. Let $\mathcal{C} \subset \mathcal{M}$ be a minimal absorbing component. The set

$$\mathcal{D}_{\mathcal{C}} = \{\tilde{x} \in \mathcal{P}_M \mid \exists n_0 = n_0(\tilde{x}) \ s.t. \ \tilde{w}^{\circ n}(\tilde{x}) \in \mathcal{C}, \forall n \geq n_0\}$$

is called the domain of attraction of $\mathcal{C}$.

Remark 4.28. If $\tilde{x}$ belongs to $\mathcal{C}$, then $\tilde{w}^{\circ n}(\tilde{x})$ belongs to $\mathcal{C}$ for any nonnegative integer n, so that every minimal absorbing component $\mathcal{C}$ is always contained in its domain of attraction, which is therefore nonempty. On the other hand, given any pixel $\tilde{x} \in \mathcal{P}_M$, there exists a nonnegative integer n_0 which represents the first time at which the orbit of $\tilde{x}$ under $\tilde{w}$ enters $\mathcal{M}$. From then on, the orbit in question will keep visiting points in $C_{\tilde{x}_{n_0}}$, the minimal absorbing component generated by $\tilde{x}_{n_0} = \tilde{w}^{\circ n_0}(\tilde{x})$, so that $\tilde{x}$ belongs to the domain of attraction of $C_{\tilde{x}_{n_0}}$. It follows that the domains of attraction of the various minimal absorbing components constitute a finite partition of the space of all pixels $\mathcal{P}_M$.

We are interested in studying the domains of attraction associated with the round-off version of a contractive transformation of the Euclidean plane into itself for two main reasons. The first motivation is technical in nature and is concerned with the derivation of sufficient conditions for the existence of a unique stationary distribution for the round-off process of Definition 4.6. This theoretical aspect will be developed in full detail in Section 4.4 (see Theorem 4.45). The second reason that has led us to devote some time to the study of domains of attraction *per se* is the surprising variety and complexity of their shapes. Examples and more detailed descriptions of these regions will be presented in Section 4.3.

We intend to conclude the present section by addressing an issue related to the periodicity of the elements that belong to the minimal absorbing set $\mathcal{M}$ of the round-off version of a contractive transformation of the Euclidean plane into itself. In order to do so, we introduce a notion of coupling between orbits.

Definition 4.29. *Assume that (A1) and (A2) hold, and let $\tilde{w}$ be the round-off version of w of accuracy $1/M$. Let $\tilde{x}$ and $\tilde{y}$ be two pixels in $\mathcal{P}_M$. We say that the two orbits under $\tilde{w}$, of $\tilde{x}$ and $\tilde{y}$, couple, if there exists a nonnegative integer n such that $\tilde{w}^{\circ n}(\tilde{x}) = \tilde{w}^{\circ n}(\tilde{y})$.*

It is clear that, if two orbits couple, they must coincide from a certain point on. Hence, if two pixels belong to the domains of attraction of two distinct minimal absorbing components, it is impossible for their two orbits to ever couple. On the other hand, if two pixels $\tilde{x}$ and $\tilde{y}$ belong to the same domain of attraction, their orbits will be absorbed by the same component, and it is possible that they will couple. More precisely, suppose that the orbit of $\tilde{x}$ first enters the minimal absorbing component at time n_1, while the orbit of $\tilde{y}$ first enters the same component at time n_2. Then, letting $n = \max(n_1, n_2)$, if $\tilde{w}^{\circ n}(\tilde{x}) = \tilde{w}^{\circ n}(\tilde{y})$, the two orbits couple. On the other hand, if $\tilde{w}^{\circ n}(\tilde{x}) \neq \tilde{w}^{\circ n}(\tilde{y})$, the fact that elements of a minimal absorbing component are periodic makes it impossible for the two orbits to ever couple.

By looking at this situation from a slightly different point of view, it is easy to see that the orbits of two distinct pixels belonging to the same minimal absorbing component will never couple. This implies that there exist as many noncoupling orbits of pixels belonging to a minimal absorbing component as there are pixels in the component itself. The orbit of any pixel belonging to the domain of attraction of a minimal absorbing component will eventually coincide with the orbit of one, and only one, pixel belonging to the minimal absorbing component. It follows that every domain of attraction can be decomposed into a finite number of subdomains which are defined as follows.

Definition 4.30. *Assume that (A1) and (A2) hold, and let $\tilde{w}$ be the round-off version of w of accuracy $1/M$. Let $C \subset \mathcal{M}$ be a minimal absorbing component, and let $\mathcal{D}_C$ be its domain of attraction. For any given pixel $\tilde{x}$ belonging to C, the subdomain of attraction of $\tilde{x}$ is defined to be the set of all pixels $\tilde{y}$ in $\mathcal{D}_C$ such that the orbit of $\tilde{y}$ under $\tilde{w}$ couples with the orbit of $\tilde{x}$ under $\tilde{w}$. Such a set is denoted by $\mathcal{D}_{\tilde{x}}$.*

It turns out that subdomains of attraction, like domains of attraction, can have varied and unusual shapes, of which we shall see some interesting examples in the following section.

4.3. Examples of Round-Off Versions of Contractive Transformations on $\mathbb{R}^2$

In this section, we shall examine the behavior of the round-off versions of some affine, contractive transformations of the Euclidean plane into itself. Every transformation w that we shall consider will be affine and will satisfy assumption (A1) of page 75. The following lemma states an important property that relates round-off versions of different accuracies of an affine, contractive transformation, whose fixed point coincides with the origin of the plane.

Lemma 4.31. *Assume that (A1) and (A2) hold, and that w is affine. Let $\tilde{w}^L$ and $\tilde{w}^M$ be round-off versions of w of accuracy $1/L$ and $1/M$, respectively. Then, $\tilde{w}^L(P_L(i,j)) = P_L(k,l)$ if and only if $\tilde{w}^M(P_M(i,j)) = P_M(k,l)$.*

Proof. The statement of the lemma is an immediate consequence of Definition 4.4 and of the fact that, since w is an affine transformation with fixed point at the origin , we must have $w(cx) = cw(x)$, for any real constant c, and any $x \in \mathbb{R}^2$.

It should be clear from the previous lemma that, when w is an affine transformation with fixed point at the origin, the level of accuracy $1/M$ only affects the dimension of the pixels. The structure and properties of the minimal absorbing set, minimal absorbing components, domains of attraction, and subdomains of attraction of $\tilde{w}$ are determined by w and are independent of M. By varying M, we can modify the level of magnification at which we look at these objects, without altering their shapes and characteristics. It is therefore without any loss of generality that we shall assume, throughout the following discussion, that the value of M equals one, so that the pixels will have edges of unit length.

One of the simplest affine transformations of the Euclidean plane into itself is the *homothety* w defined, for any $x \in \mathbb{R}^2$, by $w(x) = sx$. We shall restrict our attention to the case in which $0 \leq s < 1$. When w is applied to a point x in the plane, the point is pulled towards the origin of the coordinate system, along the segment joining it to the origin. The factor by which the distance between the point and the origin is reduced is given by s. For any two points x and y in $\mathbb{R}^2$, we have $d(w(x), w(y)) = d(sx, sy) = sd(x, y)$. Furthermore, $w((0,0)^T) = (s0, s0)^T = (0,0)^T$. Thus, w is a strict contraction whose fixed point coincides with the origin $O = (0,0)^T$ of the coordinate system, and assumptions (A1) and (A2) on page 75 are satisfied. It follows that the theory developed in the previous section applies to $\tilde{w}$, the round-off version of w of accuracy $1/M$. Notice that, in this case, s is exactly the factor by which the distance between any two points is reduced when w is applied to both of them.

For this reason, we shall refer to s as *the* contractivity factor for w.

We intend to begin the study of $\tilde{w}$ by determining what pixels belong to its minimal absorbing set $\mathcal{M}$. We first note that w maps each quadrant into itself in a similar fashion, so that we can restrict our attention to the region $\{(x_1, x_2)^T \mid x_1 \geq 0, x_2 \geq 0\}$. Let us then consider a pixel $P(i,j)$, with $i \geq 0$ and $j \geq 0$, and apply the transformation w to its center. We have $w((i,j)^T) = (si, sj)^T$. Now, if $si > i - 1/2$ and $sj > j - 1/2$, $w((i,j)^T)$ belongs to $P(i,j)$, so that $\tilde{w}(P(i,j)) = P(i,j)$. It follows that the orbit of $P(i,j)$ under $\tilde{w}$ visits $P(i,j)$ infinitely often and, by Lemma 4.17, $P(i,j)$ belongs to $\mathcal{M}$. On the other hand, suppose that $si < i - 1/2$. Then, we must have $\tilde{w}(P(i,j)) = P(i',j')$, with $0 \leq i' < i$ and $0 \leq j' \leq j$. A straightforward induction argument implies that $\tilde{w}^{\circ n}(P(i,j)) \neq P(i,j)$, for any integer $n \geq 1$. Hence, no orbit under $\tilde{w}$ will visit $P(i,j)$ infinitely often, and, by Lemma 4.17, $P(i,j)$ does not belong to $\mathcal{M}$. Similarly, if $sj < j - 1/2$, $P(i,j)$ cannot belong to $\mathcal{M}$.

Summarizing, if i and j are such that $0 \leq i < 1/(2(1-s))$ and $0 \leq j < 1/(2(1-s))$, then pixel $P(i,j)$ belongs to the minimal absorbing set $\mathcal{M}$, while, if $i > 1/(2(1-s))$ or $j > 1/(2(1-s))$, $P(i,j)$ does not belong to $\mathcal{M}$. Assuming that $1/(2(1-s))$ is not an integer, it follows by symmetry that the minimal absorbing set for $\tilde{w}$ is given by the set $\mathcal{M} = \{P(i,j) \in \mathcal{P} \mid -1/(2(1-s)) < i, j < 1/(2(1-s))\}$. Now, if $0 \leq s < 1/2$, we have $1/2 \leq 1/(2(1-s)) < 1$, and $\mathcal{M}$ reduces to the only pixel $P(0,0)$. This is in agreement with Lemma 4.26. If $1/2 < s < 3/4$, then $1 < 1/(2(1-s)) < 2$, and $\mathcal{M}$ consists of the nine pixels $P(i,j)$ such that $-1 \leq i, j \leq 1$.

The situation is a little more complicated when $s = 1/2$ (in which case $1/(2(1-s)) = 1$) because of the fact that, by Definition 4.1, a pixel contains its south and west sides, but does not contain its north and east sides. It is not difficult to verify that, in such a case, $\mathcal{M}$ consists of the four pixels $P(i,j)$, with $0 \leq i, j \leq 1$.

In general, if $(2n-1)/(2n) < s < (2n+1)/(2(n+1))$, with n a positive integer, then $\mathcal{M}$ consists of the $(2n+1)^2$ pixels $P(i,j)$, with $-n \leq i, j \leq n$. When $s = (2n-1)/(2n)$, with n a positive integer, the slight asymmetry in the definition of a pixel causes $\mathcal{M}$ to consist of the $(2n)^2$ pixels $P(i,j)$, with $-(n-1) \leq i, j \leq n$.

It is worthwhile noticing that, in this example, the condition of Lemma 4.26 is not only sufficient but also necessary, since the cardinality of $\mathcal{M}$ equals one when $0 \leq s < 1/2$, and it is strictly greater than one when $1/2 \leq s < 1$.

If we look back at the previous discussion, we see that, whenever $P(i,j)$ belongs to $\mathcal{M}$, $\tilde{w}^{\circ n}\left(P(i,j)\right) = P(i,j)$. In other words, every pixel in the minimal absorbing set has period one and, therefore, constitutes a minimal absorbing component. Thus, each pixel in $\mathcal{M}$ has an associated domain of attraction that, now, we would like to determine.

The case when $0 \leq s < 1/2$ is extremely simple, since $P(0,0)$ is the only element in $\mathcal{M}$, so that the collection of all pixels $\mathcal{P}$ constitutes its domain of attraction.

In the case when $(2n-1)/(2n) < s < (2n+1)/(2(n+1))$, with n a positive integer, we have observed that the minimal absorbing set is a square object centered at the origin, consisting of $(2n+1)^2$ pixels. We shall see that domains of attraction of pixels belonging to the outer perimeter of the square have a structure that differs significantly from that of domains of attraction of pixels in the interior of the square.

Let an integer k be given, with $k > n$. Then

$$sk < \frac{2n+1}{2(n+1)}k = \frac{2(n+1)}{2(n+1)}k - \frac{1}{2(n+1)}k \leq k - \frac{1}{2}$$

and

$$sk > \frac{2n-1}{2n}k > \frac{2n-1}{2n}n = n - \frac{1}{2},$$

so that $n - 1/2 < sk < k - 1/2$. Similarly, if $k < -n$, we have $k + 1/2 < sk < -n + 1/2$. It follows that, given any pixels $P(k,j)$ and $P(i,k)$, with $|k| > n$, $P(k',j') = \tilde{w}\left(P(k,j)\right)$ and $P(i',k') = \tilde{w}\left(P(i,k)\right)$ must be such that $n \leq |k'| < |k|$, with k' having the same sign as k. On the other hand, if k is an integer, with $0 \leq k \leq n$, then $sk < k$ and

$$sk > \frac{2n-1}{2n}k = k - \frac{1}{2n}k \geq k - 1/2,$$

so that $k - 1/2 < sk < k$. Similarly, if $-n \leq k < 0$, we have $k < sk < k + 1/2$. Hence, given any pixels $P(k,j)$ and $P(i,k)$, with $|k| \leq n$, $P(k',j') = \tilde{w}\left(P(k,j)\right)$ and $P(i',k') = \tilde{w}\left(P(i,k)\right)$ must be such that $k' = k$.

From the above, we conclude immediately that an orbit under $\tilde{w}$ will be absorbed by $P(n,n)$ if and only if it is the orbit of a pixel

$P(i, j)$ with $i \geq n$ and $j \geq n$. In other words, the domain of attraction
of $P(n, n)$ is given by the north–east quadrant of pixels with vertex
$P(n, n)$. Similarly, the domains of attraction of the other three pixels
at the corners of $\mathcal{M}$ are given by the quadrants that have such pixels
as their vertices. Just as easily, we can see that an orbit under $\tilde{w}$
will be absorbed by $P(n, n - 1)$ if and only if it is the orbit of a
pixel $P(i, n - 1)$ with $i \geq n$. In general, the domain of attraction of
a pixel that belongs to the edges of $\mathcal{M}$, but does not coincide with
a corner, is given by the infinite strip of pixels perpendicular to $\mathcal{M}$
that extends from that pixel to the outside of $\mathcal{M}$. Finally, if a pixel
$P(i, j)$ belongs to the interior of $\mathcal{M}$, i.e., $0 \leq |i|, |j| < n$, only the
orbit under $\tilde{w}$ of the pixel itself can be absorbed by it. Hence, the
domain of attraction reduces to just one pixel.

The whole situation is illustrated in Figure 4.2. Pixels on the
perimeter of $\mathcal{M}$ have domains of attraction of infinite cardinality
and constitute an insurmountable barrier that prevents orbits under
$\tilde{w}$ of pixels outside $\mathcal{M}$ from ever reaching the interior of the minimal absorbing set. Pixels in the interior of $\mathcal{M}$, on the other hand,
coincide with their domains of attraction, which therefore have cardinality one. Clearly, the domain of attraction of each minimal absorbing component is also the subdomain of attraction of the only
pixel constituting the component.

The case when $s = (2n - 1)/(2n)$, with n a positive integer, needs,
once again, to be treated separately. We have seen, for instance,
that, if $s = 1/2$, there are four minimal absorbing components given
by the four pixels $P(0, 0)$, $P(1, 0)$, $P(1, 1)$, and $P(0, 1)$. It is not
difficult to prove that, in this case, the domain of attraction of each
component is given by the quadrant, extending towards the outside
of $\mathcal{M}$, whose vertex coincides with the pixel which constitutes the
component itself. The general case does not present any significant
difference from the situations that we have encountered so far, and
we shall not deal with it specifically.

We turn now to the case in which the transformation w consists of
the composition of a contractive homothety and a rotation around
the origin. Such a transformation can be expressed in matrix form
as:

$$
w\left((x_1, x_2)^T\right) = \begin{pmatrix} s\cos\phi & -s\sin\phi \\ s\sin\phi & s\cos\phi \end{pmatrix} \begin{pmatrix} x_1 \\ x_2 \end{pmatrix}, \tag{4.2}
$$

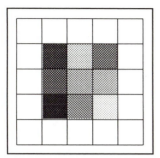

Minimal absorbing set (left) and domains of attraction (bottom) for the round-off version of the transformation w of formula (4.2), with $s = 0.6$ and $\phi = 0°$. Different shades of grey identify different minimal absorbing components, and their corresponding domains of attraction. Both images are centered at the origin.

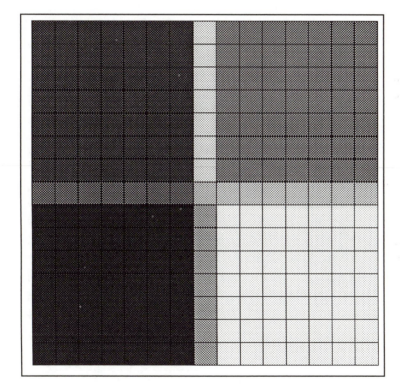

Figure 4.2. Contractive homothety.

with $0 \leq s < 1$, and $0 \leq \phi < 2\pi$. We can think of w as acting on a point $(x_1, x_2)^T$ in the plane by first contracting it towards the origin by a factor s and then rotating it around the origin (counter-

clockwise) by an angle ϕ. Alternatively, since rotations and homo-
theties commute, we can think of the rotation as being applied first
and of the contractive homothety as being applied second. It is immediately verified that w satisfies assumptions (A1) and (A2) on
page 75, so that the results outlined in the previous section apply to
its round-off version $\tilde{w}$. Here again, since s is exactly the factor by
which the distance between any two points is reduced under w, we
call s the contractivity factor for w.

We shall see that the introduction of a rotation makes it sub-
stantially more difficult to describe the behavior of $\tilde{w}$, and that, in
general, the domains of attraction of $\tilde{w}$ will present shapes that are
not easy to describe in terms of familiar geometric entities. However, in addition to the trivial case when $s < 1/2$ and the domain
of attraction of the only absorbing component coincides with the
set of all pixels, there are a few other instances in which the domains of attraction of $\tilde{w}$ assume shapes that are geometrically easy
to describe.

One such case occurs when the rotation angle ϕ equals $\pi/2$. In order to simplify the following discussion, we shall assume that the contractivity factor s is small enough for the minimal absorbing set to be
contained in the square set of nine pixels $\{P(i,j) \in \mathcal{P} \mid -1 \le i, j \le 1\}$. The argument developed in the proof of Lemma 4.11 guarantees
that a sufficient condition for this to happen is that $(\theta s)/(1-s) < 3/(2M)$, where $1/M$ denotes the accuracy of $\tilde{w}$ and $\theta = 1/(M\sqrt{2})$.
Rearranging the terms in the previous inequality, we obtain the suf-
ficient condition $s < 3/(3+\sqrt{2}) \approx 0.6796$. Since, as we have already
noted, the case when $s < 1/2$ is not particularly interesting, we shall
limit our analysis to the case when $1/2 < s < 3/(3+\sqrt{2})$.

The previous condition guarantees, as we have just said, that the
minimal absorbing set $\mathcal{M}$ for $\tilde{w}$ is contained in the set of nine pixels $\{P(i,j) \in \mathcal{P} \mid -1 \le i, j \le 1\}$. We shall now see that, when the
rotation angle ϕ equals $\pi/2$, such a set actually coincides with $\mathcal{M}$,
independently of the value of s.

Observe first that, for $\phi = \pi/2$, the transformation w takes the
form:

$$w\left((x_1, x_2)^T\right) = \begin{pmatrix} 0 & -s \\ s & 0 \end{pmatrix} \begin{pmatrix} x_1 \\ x_2 \end{pmatrix} = \begin{pmatrix} -sx_2 \\ sx_1 \end{pmatrix}.$$

Now, as is always the case, $\tilde{w}\left(P(0,0)\right) = P(0,0)$, so that $P(0,0)$ has period one with respect to $\tilde{w}$, and constitutes a minimal absorbing component of cardinality one.

If we apply the transformation w to the center of pixel $P(1,1)$, we have $w((1,1)^T) = (-s,s) \in P(-1,1)$, so that $\tilde{w}\left(P(1,1)\right) = P(-1,1)$. Similar calculations show that $\tilde{w}^{\circ 2}\left(P(1,1)\right) = P(-1,-1)$, $\tilde{w}^{\circ 3}\left(P(1,1)\right) = P(1,-1)$, and $\tilde{w}^{\circ 4}\left(P(1,1)\right) = P(1,1)$. It follows that the orbit of $P(1,1)$ under $\tilde{w}$ visits $\{P(1,1)\}$ infinitely often, so that $P(1,1)$ belongs to $\mathcal{M}$. Furthermore, $P(1,1)$ has period four and generates the minimal absorbing component $\mathcal{C}_{P(1,1)} = \{P(1,1), P(-1,1), P(-1,-1), P(1,-1)\}$.

Likewise, we have $\tilde{w}\left(P(1,0)\right) = P(0,1)$, $\tilde{w}^{\circ 2}\left(P(1,0)\right) = P(-1,0)$, $\tilde{w}^{\circ 3}\left(P(1,0)\right) = P(0,-1)$, and $\tilde{w}^{\circ 4}\left(P(1,0)\right) = P(1,0)$. Hence, $P(1,0)$ belongs to $\mathcal{M}$, has period four, and generates the minimal absorbing component $\mathcal{C}_{P(1,0)} = \{P(1,0), P(0,1), P(-1,0), P(0,-1)\}$. Since

$$\{P(0,0)\}\bigcup \mathcal{C}_{P(1,1)}\bigcup \mathcal{C}_{P(1,0)} = \{P(i,j) \in \mathcal{P} \mid -1 \leq i,j \leq 1\},$$

and $\mathcal{M}$ must be contained in $\{P(i,j) \in \mathcal{P} \mid -1 \leq i,j \leq 1\}$, it follows that $\mathcal{M}$ actually coincides with this set.

We can see that the situation is significantly different from the case in which w is a contractive homothety. The introduction of a rotational component causes some of the elements in $\mathcal{M}$ to have periods greater than one, so that there exist minimal absorbing components that do not reduce to a single pixel.

We now turn to the task of determining the domains of attraction of the three absorbing components. Consider first the quadrant of vertex $P(1,1)$, given by $\mathcal{Q}_{P(1,1)} = \{P(i,j) \mid i,j \geq 1\}$. If we apply the transformation w to the center of any pixel $P(i,j)$ belonging to $\mathcal{Q}_{P(1,1)}$, we have:

$$w\left((i,j)^T\right) = (-sj,si)^T \in \left\{(x_1,x_2)^T \in \mathbb{R}^2 \mid x_1 < -1/2, x_2 > 1/2\right\}.$$

It follows that $\tilde{w}$ maps $\mathcal{Q}_{P(1,1)}$ into the quadrant of vertex $P(-1,1)$ given by $\mathcal{Q}_{P(-1,1)} = \{P(i,j) \in \mathcal{P} \mid i \leq -1, j \geq 1\}$. It is also easy to see that every pixel in $\mathcal{Q}_{P(-1,1)}$ has at least one preimage in $\mathcal{Q}_{P(1,1)}$, so that $\tilde{w}$ maps $\mathcal{Q}_{P(1,1)}$ onto $\mathcal{Q}_{P(-1,1)}$. Similarly, letting $\mathcal{Q}_{P(-1,-1)} = \{P(i,j) \in \mathcal{P} \mid i \leq -1, j \leq -1\}$, and $\mathcal{Q}_{P(1,-1)} = \{P(i,j) \in \mathcal{P} \mid i \geq 1,$

$j \leq -1\}$, it can be readily seen that $\tilde{w}$ maps $\mathcal{Q}_{P(-1,1)}$ onto $\mathcal{Q}_{P(-1,-1)}$, $\mathcal{Q}_{P(-1,-1)}$ onto $\mathcal{Q}_{P(1,-1)}$, and $\mathcal{Q}_{P(1,-1)}$ onto $\mathcal{Q}_{P(1,1)}$.

Consider now the horizontal strip of pixels, starting at $P(1,0)$, given by $\mathcal{S}_{P(1,0)} = \{P(i,j) \in \mathcal{P} \,|\, i \geq 1, j = 0\}$. If w is applied to the center of any pixel $P(i,j)$ belonging to $\mathcal{S}_{P(1,0)}$, we have:

$$
\begin{aligned}
w\left((i,j)^T\right) &= w\left((i,0)^T\right) = (0, si)^T \\
&\in \left\{(x_1, x_2)^T \in \mathbb{R}^2 \,\middle|\, x_1 = 0, x_2 > 1/2\right\}.
\end{aligned}
$$

It follows that $\tilde{w}$ maps the strip $\mathcal{S}_{P(1,0)}$ into the vertical strip, starting at $P(0,1)$, given by $\mathcal{S}_{P(0,1)} = \{P(i,j) \in \mathcal{P} \,|\, i = 0, j \geq 1\}$. It also can easily be seen that every pixel in $\mathcal{S}_{P(0,1)}$ has at least one preimage in $\mathcal{S}_{P(1,0)}$, so that $\tilde{w}$ maps $\mathcal{S}_{P(1,0)}$ onto $\mathcal{S}_{P(0,1)}$. Likewise, $\tilde{w}$ maps $\mathcal{S}_{P(0,1)}$ onto $\mathcal{S}_{P(-1,0)} = \{P(i,j) \in \mathcal{P} \,|\, i \leq -1, j = 0\}$, $\mathcal{S}_{P(-1,0)}$ onto $\mathcal{S}_{P(0,-1)} = \{P(i,j) \in \mathcal{P} \,|\, i = 0, j \leq -1\}$, and $\mathcal{S}_{P(0,-1)}$ onto $\mathcal{S}_{P(1,0)}$. Finally, $\tilde{w}$ maps $P(0,0)$ into itself.

As an immediate consequence of this discussion, we have that the domain of attraction of $\mathcal{C}_{P(1,1)}$ is given by the union of the four quadrants $\mathcal{Q}_{P(1,1)}$, $\mathcal{Q}_{P(-1,1)}$, $\mathcal{Q}_{P(-1,-1)}$, and $\mathcal{Q}_{P(1,-1)}$, and each quadrant constitutes the subdomain of attraction associated with its vertex. Similarly, the domain of attraction of $\mathcal{C}_{P(1,0)}$ is given by the union of the four strips $\mathcal{S}_{P(1,0)}$, $\mathcal{S}_{P(0,1)}$, $\mathcal{S}_{P(-1,0)}$, and $\mathcal{S}_{P(0,-1)}$, and each strip constitutes the subdomain of attraction of its starting pixel. The domain of attraction of $P(0,0)$ reduces to $P(0,0)$ itself. It is worthwhile noting explicitly that the previous analysis holds true independently of the particular value of s between $1/2$ and $3/(3+\sqrt{2})$. We shall see that this is not the case, in general, for rotation angles different than $\pi/2$.

Figures 4.3 and 4.4 clearly show that more structure has been added to the behavior of $\tilde{w}$ by the introduction of a rotational component for w, in addition to the homothetic contractive component.

Still assuming that $1/2 < s < 3/(3+\sqrt{2})$, it is then instructive to compare the round-off version $\tilde{w}_1$ of the contractive homothety w_1 to the round-off version $\tilde{w}_2$ of the transformation w_2, which is given by the composition of w_1 with a rotation by an angle $\phi = \pi/2$ (see Figure 4.2, and Figures 4.3 and 4.4).

The minimal absorbing set for both $\tilde{w}_1$ and $\tilde{w}_2$ is given by the square set of pixels $\mathcal{M} = \{P(i,j) \in \mathcal{P} \,|\, -1 \leq i, j \leq 1\}$, and, in both

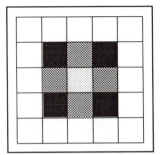

Minimal absorbing set (left) and domains of attraction (bottom) for the round-off version of the transformation w of formula (4.2), with $s = 0.6$ and $\phi = 90°$. Different shades of grey identify different minimal absorbing components, and their corresponding domains of attraction. Both images are centered at the origin.

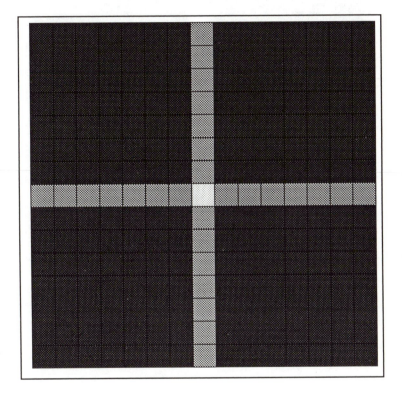

Figure 4.3. Composition of contractive homothety and rotation by a right angle (I).

cases, $P(0,0)$ constitutes a minimal absorbing component. However, while each pixel in $\mathcal{M}$ has period one with respect to $\tilde{w}_1$ and constitutes by itself a minimal absorbing component for $\tilde{w}_1$, the rotational

Subdomains of attraction (identified by different shades of grey) for the round-off version of the transformation w of formula (4.2), with $s = 0.6$ and $\phi = 90°$. The image is centered at the origin.

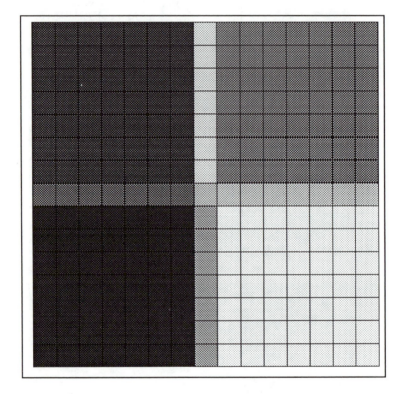

Figure 4.4. Composition of contractive homothety and rotation by a right angle (II).

component in $\tilde{w}_2$ causes the pixels on the perimeter of $\mathcal{M}$ to have period four with respect to $\tilde{w}_2$. The pixels at the corners of $\mathcal{M}$, then, group together to form the minimal absorbing component $\mathcal{C}_{P(1,1)}$, while the remaining four pixels come to form the minimal absorbing component $\mathcal{C}_{P(1,0)}$. As a final note, we observe that the domain of attraction of any pixel in $\mathcal{M}$, when the pixel is regarded as a minimal absorbing component with respect to $\tilde{w}_1$, coincides with the subdomain of attraction of the same pixel with respect to $\tilde{w}_2$.

As we have previously said, the condition that $1/2 < s < 3/(3 + \sqrt{2})$ implies that the minimal absorbing set $\mathcal{M}$ for the round-off version $\tilde{w}$ of the transformation w defined in Equation (4.2) is contained in the square set of nine pixels $\{P(i, j) \in \mathcal{P} \mid -1 \leq i, j \leq 1\}$. Which of the nine pixels actually belong to $\mathcal{M}$ and the structure of the minimal absorbing components will depend, in general, on the rotation angle ϕ and on the contractivity factor s. In the previous example, we have examined in detail the special case when $\phi = \pi/2$, noting that the results were independent of the value of s in the given range.

In the general case, the composition of $\mathcal{M}$ and its subdivision into minimal absorbing components can be similarly obtained, by following the evolution of the orbit under $\tilde{w}$ of every pixel belonging to $\{P(i, j) \in \mathcal{P} \mid -1 \leq i, j \leq 1\}$. More specifically, in light of Lemmas 4.17 and 4.19, and since $\mathcal{M}$ is contained in $\{P(i, j) \in \mathcal{P} \mid -1 \leq i, j \leq 1\}$, which has cardinality nine, a pixel $\tilde{x}$ in the latter set belongs to $\mathcal{M}$ if and only if its orbit under $\tilde{w}$ returns to it in at most nine steps. If $\tilde{x}$ belongs to $\mathcal{M}$, the actual number n of steps required for the orbit of $\tilde{x}$ to first go back to it constitutes the period of $\tilde{x}$. Finally, the minimal absorbing component generated by $\tilde{x}$ can be easily determined by applying Definition 4.22.

We shall omit the details of the computations and only present a global summary of the behavior of $\tilde{w}$ as a function of ϕ, for any fixed contractivity factor s between $1/2$ and $3/(3 + \sqrt{2})$. While, as we have just mentioned, the minimal absorbing set and the minimal absorbing components for $\tilde{w}$ can be easily determined analytically, the domains of attraction present, in general, extremely complicated shapes. Their determination by means of direct calculation would be extremely tedious. Therefore, in order to perform our analysis, we have chosen to resort to the aid of a computer program that, for any given contractivity factor and rotation angle, produces an image of the domains and subdomains of attraction of $\tilde{w}$. A brief description of the algorithm that we have employed is provided at the end of this section.

Let s then be any fixed number between $1/2$ and $3/(3 + \sqrt{2})$. We can distinguish the following cases:

Case 1: $0 < \phi < \arccos\left(\dfrac{1}{(2\sqrt{2})s}\right) - \dfrac{\pi}{4}$.

The minimal absorbing set for $\tilde{w}$ coincides with the set of nine pixels $\{P(i,j) \in \mathcal{P} \mid -1 \le i,j \le 1\}$, and each pixel in such a set constitutes a minimal absorbing component. Figure 4.5 shows the minimal absorbing set and the coinciding domains and subdomains of attraction for $\tilde{w}$ when $s = 0.6$ and $\phi = 5°$.

Case 2: $\arccos\left(\dfrac{1}{(2\sqrt{2})s}\right) - \dfrac{\pi}{4} < \phi < \arccos\left(\dfrac{1}{2s}\right)$.

The minimal absorbing set for $\tilde{w}$ is given by the five pixels $P(0,0)$, $P(1,0)$, $P(0,1)$, $P(-1,0)$, and $P(0,-1)$. Each one of these pixels constitutes a separate minimal absorbing component. The minimal absorbing set and the coinciding domains and subdomains of attraction for $\tilde{w}$ are illustrated in Figure 4.6 for the case $s = 0.6$ and $\phi = 30°$.

Case 3: $\arccos\left(\dfrac{1}{2s}\right) < \phi < \dfrac{\pi}{2} - \arccos\left(\dfrac{1}{2s}\right)$.

The minimal absorbing set for $\tilde{w}$ reduces to only the pixel $P(0,0)$.

Case 4: $\dfrac{\pi}{2} - \arccos\left(\dfrac{1}{2s}\right) < \phi < \dfrac{\pi}{2} - \left[\arccos\left(\dfrac{1}{(2\sqrt{2})s}\right) - \dfrac{\pi}{4}\right]$.

The minimal absorbing set for $\tilde{w}$ is given by the union of the two minimal absorbing components $P(0,0)$ and $\{P(1,0),\ P(0,1),\ P(-1,0),\ P(0,-1)\}$. Figures 4.7 and 4.8 show the minimal absorbing set and the domains and subdomains of attraction for $\tilde{w}$ for the case $s = 0.6$ and $\phi = 60°$.

Case 5:

$$\dfrac{\pi}{2} - \left[\arccos\left(\dfrac{1}{(2\sqrt{2})s}\right) - \dfrac{\pi}{4}\right] < \phi < \dfrac{\pi}{2} + \left[\arccos\left(\dfrac{1}{(2\sqrt{2})s}\right) - \dfrac{\pi}{4}\right].$$

The minimal absorbing set for $\tilde{w}$ is given by the union of the three minimal absorbing components $P(0,0)$, $\{P(1,0),\ P(0,1),\ P(-1,0)$,

$P(0, -1)\}$, and $\{P(1,1), P(-1,1), P(-1,-1), P(1,-1)\}$. Notice that this situation contains the special case $\phi = \pi/2$ that we have previously examined in detail. See Figures 4.9 and 4.10 for an illustration of the minimal absorbing set and of the domains and subdomains of attraction for $\tilde{w}$ when $s = 0.6$ and $\phi = 95°$. The subdomains of attraction are also illustrated in Color Plate 6.

Case 6: $\dfrac{\pi}{2} + \left[\arccos\left(\dfrac{1}{(2\sqrt{2})s}\right) - \dfrac{\pi}{4}\right] < \phi < \dfrac{\pi}{2} + \arccos\left(\dfrac{1}{2s}\right).$

The minimal absorbing set for $\tilde{w}$ is given by the union of the two minimal absorbing components $P(0,0)$ and $\{P(1,0), P(0,1), P(-1,0), P(0,-1)\}$. Figures 4.11 and 4.12 show the minimal absorbing set and the domains and subdomains of attraction for $\tilde{w}$ when $s = 0.6$ and $\phi = 120°$.

Case 7: $\dfrac{\pi}{2} + \arccos\left(\dfrac{1}{2s}\right) < \phi < \pi - \arccos\left(\dfrac{1}{2s}\right).$

The minimal absorbing set for $\tilde{w}$ reduces to $P(0,0)$.

Case 8: $\pi - \arccos\left(\dfrac{1}{2s}\right) < \phi < \pi - \left[\arccos\left(\dfrac{1}{(2\sqrt{2})s}\right) - \dfrac{\pi}{4}\right]$

The minimal absorbing set for $\tilde{w}$ is given by the union of the three minimal absorbing components $P(0,0)$, $\{P(1,0), P(-1,0)\}$, and $\{P(0,1), P(0,-1)\}$. The minimal absorbing set and the domains and subdomains of attraction for $\tilde{w}$ are shown in Figures 4.13 and 4.14 for $s = 0.6$ and $\phi = 150°$.

Case 9: $\pi - \left[\arccos\left(\dfrac{1}{(2\sqrt{2})s}\right) - \dfrac{\pi}{4}\right] < \phi < \pi.$

The minimal absorbing set for $\tilde{w}$ is given by the union of the five minimal absorbing components $P(0,0)$, $\{P(1,0), P(-1,0)\}$, $\{P(0,1), P(0,-1)\}$, $\{P(1,1), P(-1,-1)\}$, and $\{P(-1,1), P(1,-1)\}$. The minimal absorbing set and the domains and subdomains of attraction for $\tilde{w}$ are illustrated in Figures 4.15 and 4.16 for $s = 0.6$ and $\phi = 175°$.

The analysis for rotations between π and 2π can be obtained by symmetry from the above. Once again, we have neglected to consider some cases in which the behavior of $\tilde{w}$ is affected by the fact that pixels have been defined to be half open. Such cases, however, do not present any peculiar feature and can be easily treated separately.

In general, given a positive integer k, a sufficient condition for the minimal absorbing set to be contained in the square set of $(2k+1)^2$ pixels centered at the origin is that $s < (2k+1)/(2k+1+\sqrt{2})$ (see the proof of Lemma 4.11 for a straightforward justification of this condition). This fact and the same type of arguments developed in the previous discussion can be employed to analyze the behavior of $\tilde{w}$ for any value of the contractivity factor s and of the rotation angle ϕ. Instead of doing so systematically, we prefer to conclude our treatment of round-off versions of transformations of the type defined by Equation (4.2) by presenting three more examples that we find interesting.

The first example deals with a transformation w whose contractivity factor s equals $(0.51)\sqrt{2} \approx 0.7212$, and whose rotation angle ϕ equals $\pi/4$. In matrix form, the transformation w is given by:

$$w\left((x_1, x_2)^T\right) = \begin{pmatrix} 0.51 & -0.51 \\ 0.51 & 0.51 \end{pmatrix} \begin{pmatrix} x_1 \\ x_2 \end{pmatrix}.$$

If the contractivity factor had been chosen to be between $1/2$ and $3/(3+\sqrt{2})$, we would be in the situation previously denoted by Case 3, and the minimal absorbing set for $\tilde{w}$, the round-off version of w, would reduce to the only pixel $P(0,0)$. In the present example, though, the contractivity factor of w is larger than $3/(3+\sqrt{2})$, and this condition alone is enough to give rise to a situation far more interesting. The minimal absorbing set $\mathcal{M}$ for $\tilde{w}$ is given by the union of the three minimal absorbing components $P(0,0)$, $\{P(1,0),\ P(1,1),\ P(0,1),\ P(-1,1),\ P(-1,0),\ P(-1,-1),\ P(0,-1),\ P(1,-1)\}$, and $\{P(2,1),\ P(1,2),\ P(-1,2),\ P(-2,1),\ P(-2,-1),\ P(-1,-2),\ P(1,-2),\ P(2,-1)\}$.

Referring to Figure 4.17, we can see that the domain of attraction of the first component reduces to the pixel $P(0,0)$ itself. The domain of attraction of the second component is given by the union of eight strips of pixels, each of which starts at a pixel belonging to the component and extends (horizontally, vertically, or diagonally) towards the outside of $\mathcal{M}$. Finally, the domain of attraction of the third component is given by the union of the eight triangular regions whose vertices coincide with one of the pixels in the component, and that are separated one from the other by the previously mentioned strips.

It should be clear, by now, that this orderly type of structure is brought along by the choice of the rotation angle. The present situation should be compared to the case when $\phi = \pi/2$, which is illustrated in Figure 4.3, observing that each quadrant in the domain of attraction of $\mathcal{C}_{P(1,1)}$ (see page 95) has now been subdivided into two triangular regions. (Note that even though the case $\phi = \pi/2$ has been previously analyzed only for $1/2 < s < 3/(3 + \sqrt{2})$, the behavior of $\tilde{w}$, as far as its minimal absorbing set and domains of attraction are concerned, remains the same even when $s \approx 0.7212$.)

The second example illustrates the minimal absorbing set and domains of attraction for the round-off version of a contractive homothety, with contractivity factor $s = 0.9$, composed with a rotation by an angle $\phi = \pi/6$. Figures 4.18 and 4.19 show that the minimal absorbing set is given by the union of seven components, and that only two of them have domains of attraction of infinite cardinality. Perhaps more interestingly, of these two components, the one drawn in a darker shade of grey has a domain of attraction which is given by the union of *disconnected* regions. This type of behavior had not been encountered in any of the previous examples.

The last example of a transformation given by the composition of a rotation and a contractive homothety that we would like to present deals with the case $\phi = \pi/6$ and $s = 0.95$, and is illustrated in Figures 4.20 and 4.21. The contractivity factor here is considerably larger than in all the cases that we have previously considered. This fact is reflected in the large cardinality of the minimal absorbing set for $\tilde{w}$ and in the large number of minimal absorbing components. This example presents an interesting behavior, in that all but one of the 11 absorbing components have domains of attraction of finite cardinality. In other words, all but a finite number of orbits under $\tilde{w}$ will eventually belong to the same absorbing component.

In the present section, we have tried to provide a feeling for the behavior of the round-off version of a contractive transformation of the Euclidean plane into itself. In our attempt to do so, we have limited our discussion to a few, very special types of contractive transformations, consisting of either a contractive homothety or of the composition of a contractive homothety and a rotation around the origin. We have seen that, despite the inherent simplicity of such transformations, the behavior of their corresponding round-off versions can be surprisingly varied and intricate. In particular, we have noticed with interest the wide variety of shapes that domains of attractions can assume.

Our analysis is obviously far from exhaustive, its only purpose being that of providing a few, insightful examples. Experience has taught us that new, fascinating discoveries can be made each time a new transformation is analyzed, provided its smallest contractivity factor is sufficiently large. As previously mentioned, we have found it convenient to rely on the help of a computer program to conduct our analysis and produce images of the domains of attraction. The program can deal with a generic affine contraction, whose fixed point coincides with the origin. Figure 4.22, for instance, illustrates the minimal absorbing set and the domains of attraction for the round-off version of:

$$w\left((x_1, x_2)^T\right) = \begin{pmatrix} 0.5 & 0.3 \\ -0.1 & 0.4 \end{pmatrix} \begin{pmatrix} x_1 \\ x_2 \end{pmatrix}, \qquad (4.3)$$

an affine transformation that is more general than those previously studied.

The algorithm implemented in the program is based on the theoretical results presented in Section 4.2 and consists of the following steps.

1. Consider a unit viewing window centered at the origin.

2. Based on the argument developed in the proof of Lemma 4.11, choose a pixel size small enough to guarantee that the minimal absorbing set for the round-off version $\tilde{w}$ of the given contractive affine transformation w is contained in the viewing window.

3. Let $\tilde{x}$ be the pixel at the lower left corner of the viewing window. Follow its orbit under $\tilde{w}$ until the first time m at which the orbit visits a pixel that it has already visited, at time, e.g., $m - n$. Set $K = 1$, $\mathcal{C}_1 = \{\tilde{w}^{\circ(m-n)}(\tilde{x}), \ldots, \tilde{w}^{\circ(m-1)}(\tilde{x})\}$, $\mathcal{M} = \mathcal{C}_1$, and $\mathcal{D}_1 = \{\tilde{x}\}$. (Throughout the description of this algorithm, K denotes the number of minimal absorbing components $\mathcal{C}_k$, with

respective domains of attraction $\mathcal{D}_k$, that have been identified up to the current point.)

4. Let $\tilde{x}$ be a pixel in the viewing window which has not yet been considered. Follow its orbit under $\tilde{w}$ until the first time m at which either:

   (a) the orbit enters $\mathcal{M}$,

   or

   (b) the orbit visits a pixel that it has already visited, at time, e.g., $m - n$.

   If (a) happens first, determine the minimal absorbing component $\mathcal{C}_k$, $1 \le k \le K$, to which $\tilde{w}^{\circ(m)}(\tilde{x})$ belongs and add $\tilde{x}$ to $\mathcal{D}_k$.
   If (b) happens first, increase K by 1, set $\mathcal{C}_K = \{\tilde{w}^{\circ(m-n)}(\tilde{x}), \ldots, \tilde{w}^{\circ(m-1)}(\tilde{x})\}$, $\mathcal{D}_K = \{\tilde{x}\}$, and add $\mathcal{C}_K$ to $\mathcal{M}$.

5. Repeat step 4 until all pixels in the viewing window have been considered.

6. Color-code the K minimal absorbing components and respective domains of attraction thus determined, and display them.

Extension of these ideas to the case of an algorithm that identifies and displays the various subdomains of attraction is clearly straightforward.

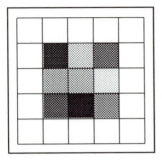

Minimal absorbing set (left) and domains of attraction (bottom) for the
round-off version of the transformation w of formula (4.2), with $s = 0.6$
and $\phi = 5°$. Different shades of grey
identify different minimal absorbing
components, and their corresponding domains of attraction. Both images are centered at the origin. The
size of the bottom image is 511×511
pixels.

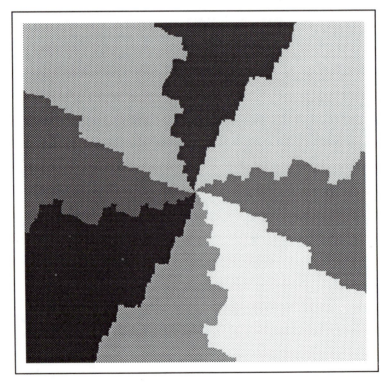

Figure 4.5. Case 1.

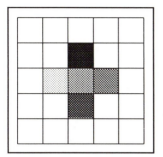

Minimal absorbing set (left) and domains of attraction (bottom) for the round-off version of the transformation w of formula (4.2), with $s = 0.6$ and $\phi = 30°$. Different shades of grey identify different minimal absorbing components, and their corresponding domains of attraction. Both images are centered at the origin. The size of the bottom image is 511×511 pixels.

Figure 4.6. Case 2.

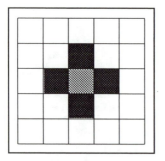

Minimal absorbing set (left) and domains of attraction (bottom) for the round-off version of the transformation w of formula (4.2), with $s = 0.6$ and $\phi = 60°$. Different shades of grey identify different minimal absorbing components, and their corresponding domains of attraction. Both images are centered at the origin.

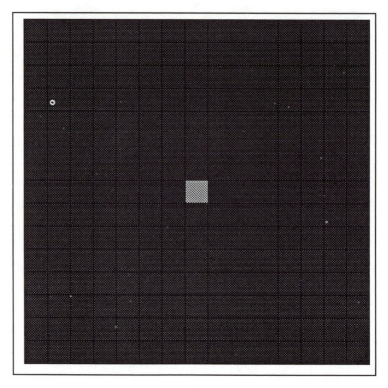

Figure 4.7. Case 4–I.

Subdomains of attraction (identified by different shades of grey) for the round-off version of the transformation w of formula (4.2), with $s = 0.6$ and $\phi = 60°$. The image is centered at the origin. Its size is 511×511 pixels.

Figure 4.8. Case 4–II.

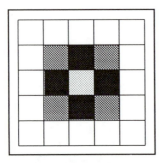

Minimal absorbing set (left) and domains of attraction (bottom) for the
round-off version of the transformation w of formula (4.2), with $s = 0.6$
and $\phi = 95°$. Different shades of
grey identify different minimal absorbing components, and their corresponding domains of attraction.
Both images are centered at the origin. The size of the bottom image
is 511×511 pixels.

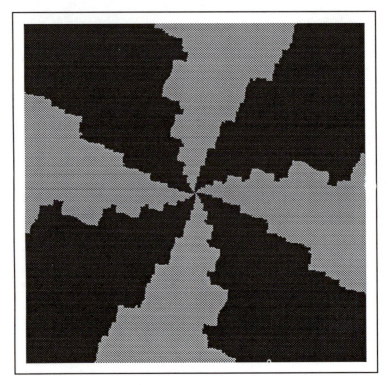

Figure 4.9. Case 5–I.

Subdomains of attraction (identified by different shades of grey) for the round-off version of the transformation w of formula (4.2), with $s = 0.6$ and $\phi = 95°$. The image is centered at the origin. Its size is 511×511 pixels.

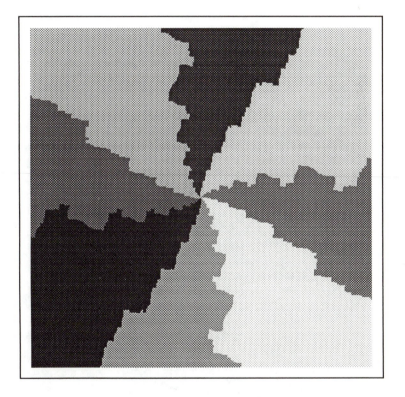

Figure 4.10. Case 5–II.

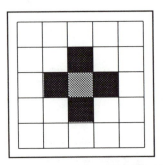

Minimal absorbing set (left) and
domains of attraction (bottom)
for the round-off version of the
transformation w of formula (4.2),
with $s = 0.6$ and $\phi = 120°$.
Different shades of grey identify
different minimal absorbing components, and their corresponding
domains of attraction. Both images are centered at the origin.

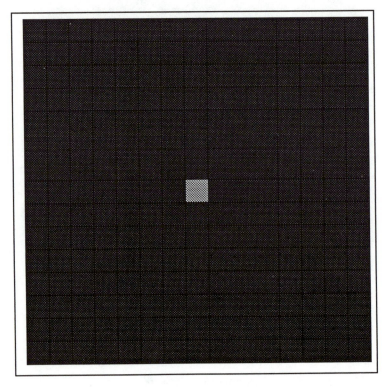

Figure 4.11. Case 6–I.

Subdomains of attraction (identified by different shades of grey) for the round-off version of the transformation w of formula (4.2), with $s = 0.6$ and $\phi = 120°$. The image is centered at the origin. Its size is 511×511 pixels.

Figure 4.12. Case 6–II.

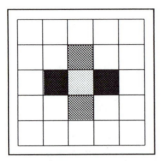

Minimal absorbing set (left) and domains of attraction (bottom) for the round-off version of the transformation w of formula (4.2), with $s = 0.6$ and $\phi = 150°$. Different shades of grey identify different minimal absorbing components, and their corresponding domains of attraction. Both images are centered at the origin. The size of the bottom image is 511×511 pixels.

Figure 4.13. Case 8–I.

Subdomains of attraction (identified by different shades of grey) for the round-off version of the transformation w of formula (4.2), with $s = 0.6$ and $\phi = 150°$. The image is centered at the origin. Its size is 511×511 pixels.

Figure 4.14. Case 8–II.

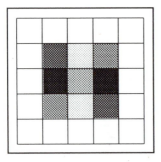

Minimal absorbing set (left) and domains of attraction (bottom) for the
round-off version of the transformation w of formula (4.2), with $s = 0.6$
and $\phi = 175°$. Different shades of
grey identify different minimal absorbing components, and their corresponding domains of attraction.
Both images are centered at the origin. The size of the bottom image
is 511×511 pixels.

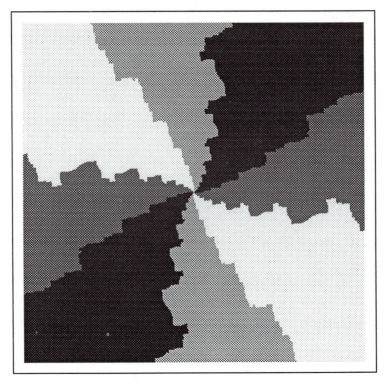

Figure 4.15. Case 9–I.

Subdomains of attraction (identified by different shades of grey) for the round-off version of the transformation w of formula (4.2), with $s = 0.6$ and $\phi = 175°$. The image is centered at the origin. Its size is 511×511 pixels.

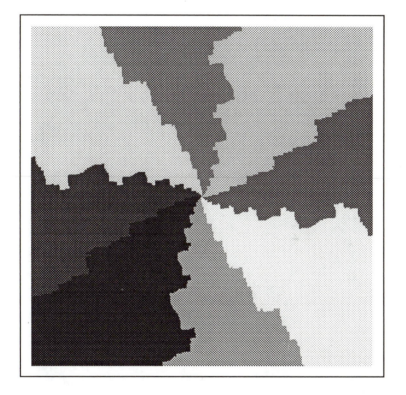

Figure 4.16. Case 9–II.

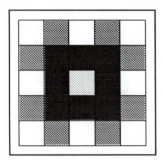

Minimal absorbing set (left) and domains of attraction (bottom) for the round-off version of the transformation w of formula (4.2), with $s = 0.7212$ and $\phi = 45°$. Different shades of grey identify different minimal absorbing components, and their corresponding domains of attraction. Both images are centered at the origin.

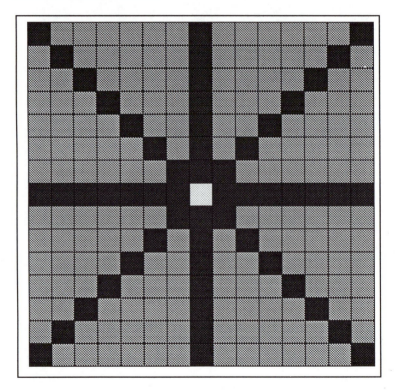

Figure 4.17. Effect of larger contractivity factor (I).

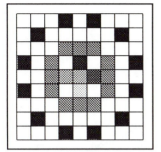

Minimal absorbing set (left) and domains of attraction (bottom) for the round-off version of the transformation w of formula (4.2), with $s = 0.9$ and $\phi = 30°$. Different shades of grey identify different minimal absorbing components, and their corresponding domains of attraction. Both images are centered at the origin.

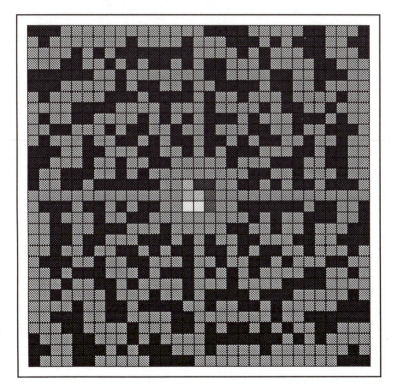

Figure 4.18. Effect of larger contractivity factor (II–a).

Domains of attraction (identified by different shades of grey) for the round-off version of the transformation w of formula (4.2), with $s = 0.9$ and $\phi = 30°$. The image is centered at the origin. Its size is 511×511 pixels.

Figure 4.19. Effect of larger contractivity factor (II–b).

Minimal absorbing set for the round-off version of the transformation w of formula (4.2), with $s = 0.95$ and $\phi = 30°$. Different shades of grey identify different minimal absorbing components. The image is centered at the origin.

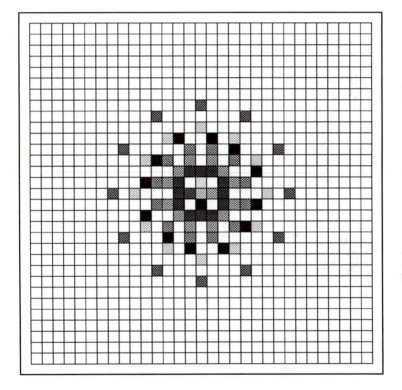

Figure 4.20. Effect of larger contractivity factor (III–a).

Domains of attraction for the round-off version of the
transformation w of formula (4.2), with $s = 0.95$ and
$\phi = 30°$. Different shades of grey identify different domains of attraction corresponding to the minimal absorbing components illustrated in Figure 4.20. The
image is centered at the origin.

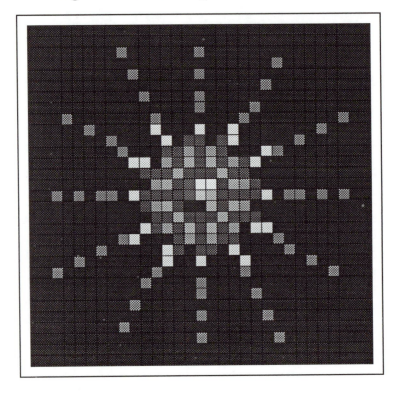

Figure 4.21. Effect of larger contractivity factor (III–b).

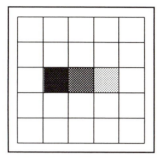

Minimal absorbing set (left) and domains of attraction (bottom) for the round-off version of the transformation w of formula (4.3). Different shades of grey identify different minimal absorbing components, and their corresponding domains of attraction. Both images are centered at the origin. The size of the bottom image is 511 × 511 pixels.

Figure 4.22. General affine transformation.

4.4. Limiting Properties of the Round-Off Process

In the current section, we shall present results relating to the limiting behavior of the round-off process introduced in Definition 4.6, as the pixel size approaches zero. Mainly, we shall be concerned with determining how good an approximation to the original process defined in Chapter 3 is provided by the round-off process.

Throughout the remainder of the section, we shall make use of the assumptions that:

(A3) $w_1, \ldots, w_N$ are strict contractions, with respective contractivity factors $s_1, \ldots, s_N$, and $\{\mathbb{R}^2; w_1, \ldots, w_N; p_1, \ldots, p_N\}$ is a hyperbolic IFS with contractivity factor $s = \max_{1 \leq n \leq N} s_n$, whose associated Markov chain and round-off process of accuracy $1/M$ are denoted by $\{X_n\}_{n=0}^{\infty}$ and $\{\tilde{X}_n\}_{n=0}^{\infty}$ respectively.

(A4) The fixed point of w_1 coincides with $O = (0,0)^T$.

The latter assumption does not cause any substantial loss of generality, since the condition can always be met upon translation of the coordinate system. By relabeling the transformations, we also can always assume that:

(A5) $s_1 = \min_{1 \leq n \leq N} s_n$.

As a first step in the direction of comparing the round-off process $\{\tilde{X}_n\}_{n=0}^{\infty}$ of accuracy $1/M$ to the original process $\{X_n\}_{n=0}^{\infty}$, we establish the following lemma that provides a bound on the distance between corresponding orbits of the two processes.

Lemma 4.32. *Assume that (A3) is satisfied. Let $\tilde{x}_0 \in \mathcal{P}_M$ be given and let $x_0 = \tilde{x}_0$. Consider two orbits $\{x_{n+1} = w_{\sigma_n}(x_n)\}_{n=0}^{\infty}$ and $\{\tilde{x}_{n+1} = \tilde{w}_{\sigma_n}(\tilde{x}_n)\}_{n=0}^{\infty}$ of $\{X_n\}_{n=0}^{\infty}$ and $\{\tilde{X}_n\}_{n=0}^{\infty}$, respectively, where, for each n, σ_n is an integer between 1 and N. Then:*

$$d(x_n, \tilde{x}_n) \leq \frac{1 - s^n}{1 - s}\theta, \quad for \; n = 1, 2, 3, \ldots, \tag{4.4}$$

where $1/M$ is the level of accuracy of $\{\tilde{X}_n\}_{n=0}^{\infty}$, and $\theta = 1/(M\sqrt{2})$ is half the length of the pixel diagonal.

Proof. The result is easily proved by means of a standard induction argument. For $n = 1$, applying the triangle inequality, we have:

$$
\begin{aligned}
d\left(x_1, \tilde{x}_1\right) &\leq d\left(x_1, w_{\sigma_0}\left(\tilde{x}_0\right)\right) + d\left(w_{\sigma_0}\left(\tilde{x}_0\right), \tilde{x}_1\right) \\
&\leq 0 + \theta \\
&= \frac{1-s}{1-s}\theta,
\end{aligned}
$$

so that inequality (4.4) holds. Here, we have made use of the fact that, by assumption, $x_1 = w_{\sigma_0}(x_0) = w_{\sigma_0}\left(\tilde{x}_0\right)$, and that $\tilde{x}_1$ is the center of the pixel to which $w_{\sigma_0}\left(\tilde{x}_0\right)$ belongs. In a similar manner, assuming that inequality (4.4) is true for n, and making additional use of the assumption that the IFS is hyperbolic with contractivity factor s, we have:

$$
\begin{aligned}
d\left(x_{n+1}, \tilde{x}_{n+1}\right) &\leq d\left(x_{n+1}, w_{\sigma_n}\left(\tilde{x}_n\right)\right) + d\left(w_{\sigma_n}\left(\tilde{x}_n\right), \tilde{x}_{n+1}\right) \\
&\leq d\left(w_{\sigma_n}(x_n), w_{\sigma_n}\left(\tilde{x}_n\right)\right) + \theta \\
&\leq sd(x_n, \tilde{x}_n) + \theta \\
&\leq s\frac{1-s^n}{1-s}\theta + \theta \\
&= \frac{s - s^{n+1} + 1 - s}{1-s}\theta \\
&= \frac{1 - s^{n+1}}{1-s}\theta,
\end{aligned}
$$

so that inequality (4.4) holds true for $n+1$, and the result follows.

The following corollary is an immediate consequence.

Corollary 4.33. *Under the assumptions of Lemma 4.32,*

$$
d\left(x_n, \tilde{x}_n\right) \leq \frac{\theta}{1-s}, \qquad for\ n = 0, 1, 2, \dots\ .
$$

Remark 4.34. The previous corollary provides a formal statement of the assertion that corresponding orbits of the two processes remain close together as they evolve over time.

We shall now make use of Corollary 4.33 to derive a fundamental property of the Markov chain $\{\tilde{X}_n\}_{n=0}^{\infty}$, which is stated in the following lemma.

Lemma 4.35. *Assume that (A3) is satisfied. Then, the set $\mathcal{R}$ of recurrent states for $\{\tilde{X}_n\}_{n=0}^{\infty}$ is finite (and nonempty). Furthermore, for any initial state, the probability of ultimate absorption in $\mathcal{R}$ is one.*

Proof. Let $x \in \mathbb{R}^2$ be given, and let $Z = \max_{1 \le i \le N} d(w_i(O), O)$. We claim that, for any orbit $\{x_n\}_{n=0}^{\infty}$ of $\{X_n\}_{n=0}^{\infty}$ starting at $x_0 = x$, we have:

$$d(x_n, O) \le s^n d(x, O) + \frac{1 - s^n}{1 - s} Z, \qquad \text{for } n = 0, 1, 2, \ldots \quad . \qquad (4.5)$$

We shall prove inequality (4.5) by induction. For $n = 0$, we have $d(x_0, O) = d(x, O)$, and inequality (4.5) holds. Suppose that inequality (4.5) is true for n. Then:

$$
\begin{aligned}
d(x_{n+1}, O) &\le d\left(w_{\sigma_n}(x_n), w_{\sigma_n}(O)\right) + d\left(w_{\sigma_n}(O), O\right) \\
&\le s d(x_n, O) + Z \\
&\le s\left(s^n d(x, O) + \frac{1 - s^n}{1 - s} Z\right) + Z \\
&= s^{n+1} d(x, O) + \frac{s - s^{n+1} + 1 - s}{1 - s} Z \\
&= s^{n+1} d(x, O) + \frac{1 - s^{n+1}}{1 - s} Z,
\end{aligned}
$$

so that inequality (4.5) is also true for $n + 1$.

Let $\epsilon > 0$ now be given. Since $\lim_{n \to \infty} s^n d(x, O) = 0$, and $((1 - s^n)/(1-s))Z < Z/(1-s)$ for all n, it follows from inequality (4.5) that there exists an integer $\bar{n} = \bar{n}(x, \epsilon)$ such that $d(x_n, O) \le Z/(1-s) + \epsilon$, for any $n \ge \bar{n}$. Define:

$$B = \left\{\tilde{y} \in \mathcal{P}_M \,\middle|\, d(\tilde{y}, O) \le \frac{\theta + Z}{1 - s} + \theta\right\},$$

where $\theta = 1/(M\sqrt{2})$. Every pixel belonging to $\mathcal{P}_M \setminus B$ is a transient state for the round-off process $\{\tilde{X}_n\}_{n=0}^{\infty}$ of accuracy $1/M$. In fact, for a given $\tilde{x} \in \mathcal{P}_M$, letting $\epsilon = \theta$ and applying Corollary 4.33, we have that there exists an integer $\bar{n} = \bar{n}(\tilde{x}, \theta)$ such that, for all orbits of $\{\tilde{X}_n\}_{n=0}^{\infty}$ starting at $\tilde{X}_0 = \tilde{x}$, and all $n \geq \bar{n}$:

$$
\begin{aligned}
d(\tilde{x}_n, O) &\leq d(\tilde{x}_n, x_n) + d(x_n, O) \\
&\leq \frac{\theta}{1-s} + \frac{Z}{1-s} + \theta \\
&= \frac{\theta + Z}{1-s} + \theta.
\end{aligned}
$$

Thus, $\tilde{x}_n$ belongs to $\mathcal{B}$ for any $n \geq \bar{n}$, so that, if $\tilde{x} \in \mathcal{P}_M \setminus \mathcal{B}$, no orbit of $\{\tilde{X}_n\}_{n=0}^{\infty}$ starting at $\tilde{x}$ will ever return to it after time $\bar{n}$. Since, for any initial state, every orbit of $\{\tilde{X}_n\}_{n=0}^{\infty}$ will eventually belong to $\mathcal{B}$ after a finite number of steps, we can apply a standard argument about Markov chains with finite state space (see Ross (1983), page 106), to conclude that the set $\mathcal{R}$ of recurrent states is nonempty and finite, being contained in $\mathcal{B}$. Furthermore, for any initial state, the probability of ultimate absorption in $\mathcal{R}$ is one.

The importance of the previous lemma lies in the fact that, since the set $\mathcal{R}$ of recurrent states is finite, and the probability of ultimate absorption in $\mathcal{R}$ is one for any initial state, we can investigate the limiting properties of $\{\tilde{X}_n\}_{n=0}^{\infty}$ using results pertaining to Markov chains with finite state space. We refer the reader to Chapter XV of Feller (1968) and Chapter 4 of Ross (1983) for a careful statement and development of the related theory.

The first consequence of $\mathcal{R}$ being finite is that every state in $\mathcal{R}$ is *positive* recurrent, i.e., its mean recurrence time is finite. This means that not only will the process $\{\tilde{X}_n\}_{n=0}^{\infty}$ return to some pixel in $\mathcal{R}$ infinitely often with probability one, but also that the expected time between successive visits is finite. Furthermore, $\mathcal{R}$ can be partitioned into the union of a finite number of nonempty, irreducible closed sets of communicating states. Denoting by $\mathcal{R}_k$ the kth irreducible closed set, and letting $K(M)$ be the total number of such sets, we can then write:

$$
\mathcal{R} = \mathcal{R}_1 \bigcup \mathcal{R}_2 \bigcup \cdots \bigcup \mathcal{R}_{K(M)},
$$

where the union is over disjoint sets.

The total number $K(M)$ of irreducible closed sets depends, in general, on the accuracy $1/M$ of the process $\{\tilde{X}_n\}_{n=0}^{\infty}$, but cannot exceed, as the following lemma shows, the number of minimal absorbing components for $\tilde{w}_1$.

Lemma 4.36. *Suppose that (A3) and (A4) are satisfied. Then, the total number $K(M)$ of irreducible closed sets for $\{\tilde{X}_n\}_{n=0}^{\infty}$, the round-off process of accuracy $1/M$, cannot exceed the number of minimal absorbing components for $\tilde{w}_1$.*

Proof. Recall, from Section 4.2, that the number of minimal absorbing components for $\tilde{w}_1$ is finite. Let $\mathcal{M}$ be the minimal absorbing set for $\tilde{w}_1$. Let a pixel $\tilde{y} \in \mathcal{M}$ be a recurrent state for $\{\tilde{X}_n\}_{n=0}^{\infty}$, and let $\mathcal{C}$ be the minimal absorbing component to which it belongs. It follows immediately from Lemma 4.19 and Definition 4.22 that all pixels in $\mathcal{C}$ communicate with $\tilde{y}$, so that they are all recurrent and they all belong to the same irreducible closed set. In other words, if a minimal absorbing component for $\tilde{w}_1$ contains a recurrent pixel, the whole component is contained in the same irreducible closed set. To prove that the statement of the lemma holds, it is therefore enough to show that every pixel in $\mathcal{R}$ (the set of recurrent states for $\{\tilde{X}_n\}_{n=0}^{\infty}$) communicates with a pixel belonging to one of the minimal absorbing components for $\tilde{w}_1$.

Let $\tilde{x} \in \mathcal{R}$ be given. If $\tilde{x}$ belongs to $\mathcal{M}$, there is nothing to prove. If $\tilde{x}$ is not in $\mathcal{M}$, it follows from the definition of an absorbing set that there exists an integer $m > 0$ such that the orbit of $\tilde{x}$ under $\tilde{w}_1$ will first enter $\mathcal{M}$ at time m. Thus, there exists a path that leads from $\tilde{x}$ to $\tilde{w}_1^{\circ m}(\tilde{x})$ with positive probability p_1^m. Suppose there existed no path leading back from $\tilde{w}_1^{\circ m}(\tilde{x})$ to $\tilde{x}$. Then, the chain $\{\tilde{X}_n\}_{n=0}^{\infty}$, starting at $\tilde{x}$, would have a positive probability of at least p_1^m of never returning to $\tilde{x}$, which contradicts the hypothesis that $\tilde{x}$ is recurrent. It follows that $\tilde{x}$ communicates with $\tilde{w}_1^{\circ m}(\tilde{x})$, and, therefore, with all the pixels in the minimal absorbing component $\mathcal{C}$ to which $\tilde{w}_1^{\circ m}(\tilde{x})$ belongs. Hence, all pixels in $\mathcal{C}$ are recurrent, since they communicate with a recurrent state, and $\tilde{x}$ belongs to the same irreducible closed set that contains $\mathcal{C}$. This completes the proof of the lemma. $\blacksquare$

Remark 4.37. It follows from the previous lemma and Remark 4.24 that $K(M)$ can be bounded above, independently of the level of accuracy $1/M$, by the integer $K(s_1)$ defined in Corollary 4.12. Furthermore, if w_1 is affine, we can replace such a bound with the number of minimal absorbing components for $\tilde{w}_1$, which is usually smaller than $K(s_1)$, and, as a consequence of Lemma 4.31, also independent of the level of accuracy $1/M$.

Let us now restrict our attention to $\mathcal{R}_k$, one of the $K(M)$ irreducible closed sets for the round-off process $\{\tilde{X}_n\}_{n=0}^{\infty}$ of accuracy $1/M$. It is a well-known result in Markov chain theory that the fact that the states in $\mathcal{R}_k$ are positive recurrent implies the existence of a unique stationary distribution π_k for $\{\tilde{X}_n\}_{n=0}^{\infty}$ on $\mathcal{R}_k$. Furthermore, a law of large numbers similar to Theorem 3.3 on page 58 guarantees that there is almost sure convergence of the empirical distributions of orbits of $\{\tilde{X}_n\}_{n=0}^{\infty}$ starting in $\mathcal{R}_k$ to the stationary distribution π_k supported on $\mathcal{R}_k$ (see the proof of the following theorem for further details). This situation is intriguing, since, if $K(M)$ is greater than one, and the algorithm based on the round-off process is used to generate digitized images, one would obtain, depending on the starting point, $K(M)$ different images corresponding to the same IFS.

Two questions arise naturally at this point. Are there any situations in which $K(M)$ is strictly greater than one? If so, how similar are the different images that one would obtain at a given level of accuracy $1/M$, and how do they relate to the unique image associated with the IFS through the original process $\{X_n\}_{n=0}^{\infty}$? We shall see in the sequel that the answer to the first question is affirmative, and we shall provide both examples of round-off processes with multiple stationary distributions and conditions for uniqueness. Bearing this in mind, we turn first to answering the second question in detail.

In order to do so, for any integer $M \geq 1$, we arbitrarily choose a distribution $\pi_{k(M)}$ among the $K(M)$ stationary distributions for the round-off process $\{\tilde{X}_n\}_{n=0}^{\infty}$ of accuracy $1/M$, and construct a sequence $\{\pi_{k(M)}\}_{M=1}^{\infty}$. The following theorem then holds.

Theorem 4.38. *Suppose that (A3) and (A4) are satisfied, and let $\{\pi_{k(M)}\}_{M=1}^{\infty}$ be a sequence of stationary distributions for the round-off processes $\{\tilde{X}_n^M\}_{n=0}^{\infty}$ of accuracy $1/M$, constructed as above. Then:*

$$\sum_{\tilde{x} \in \mathcal{R}_{k(M)}} \pi_{k(M)} \delta_{\tilde{x}} \xrightarrow{w} \nu, \qquad as\ M \to \infty,$$

where, for any Borel subset B of $\mathbb{R}^2$, $\delta_{\tilde{x}}(B)$ equals 1 if $\tilde{x} \in B$, and 0 otherwise. In other words, $\pi_{k(M)}$ converges weakly, as M goes to infinity, to the unique stationary distribution ν of the original process $\{X_n\}_{n=0}^{\infty}$.

Proof.   It is enough to show that, for any continuous function $f : \mathbb{R}^2 \longmapsto \mathbb{R}$ with compact support,

$$\lim_{M \to \infty} \sum_{\tilde{x} \in \mathcal{R}_{k(M)}} f(\tilde{x}) \, \pi_{k(M)}(\tilde{x}) = \int_{\mathbb{R}^2} f(x) \, d\nu(x),$$

where $\mathcal{R}_{k(M)}$ is the support of $\pi_{k(M)}$. We observe that, by Elton's theorem (Theorem 3.3 on page 58), letting $X_0 = x_0$ for an arbitrary $x_0 \in \mathbb{R}^2$, we have:

$$\lim_{n \to \infty} \frac{1}{n+1} \sum_{j=0}^{n} f(X_j) = \int_{\mathbb{R}^2} f(x) \, d\nu(x) \quad \text{a.s.}$$

Let M be fixed and let $\tilde{x}_0^M \in \mathcal{R}_{k(M)}$ be given. Since the cardinality of $\mathcal{R}_{k(M)}$ is finite, the conditions of Theorem 6.2 on page 220 in Doob (1953) are satisfied and, letting $\tilde{X}_0^M = \tilde{x}_0^M$, we have:

$$\lim_{n \to \infty} \frac{1}{n+1} \sum_{j=0}^{n} f\left(\tilde{X}_j^M\right) = \sum_{\tilde{x} \in \mathcal{R}_{k(M)}} f(\tilde{x}) \, \pi_{k(M)}(\tilde{x}) \quad \text{a.s.}$$

Hence, letting $\tilde{X}_0^M = X_0^M = \tilde{x}_0^M$ for an arbitrary $\tilde{x}_0^M \in \mathcal{R}_{k(M)}$, we have:

$$\left| \sum_{\tilde{x} \in \mathcal{R}_{k(M)}} f(\tilde{x}) \, \pi_{k(M)}(\tilde{x}) - \int_{\mathbb{R}^2} f(x) \, d\nu(x) \right| =$$

$$\lim_{n \to \infty} \frac{1}{n+1} \left| \sum_{j=0}^{n} \left(f\left(\tilde{X}_j^M\right) - f\left(X_j^M\right) \right) \right| \quad \text{a.s.},$$

where the superscript M in X_j^M indicates the fact that the starting point of the sequence depends on the level of accuracy. Notice that, since M varies over a countable set, the previous equality holds with probability one for all M, provided that the common starting point for both $\{X_n^M\}_{n=0}^\infty$ and $\{\tilde{X}_n^M\}_{n=0}^\infty$ is always chosen in $\mathcal{R}_{k(M)}$. Applying the triangle inequality to the right-hand side of the previous equality, we have:

$$\lim_{n\to\infty} \frac{1}{n+1} \left| \sum_{j=0}^n \left(f\left(\tilde{X}_j^M\right) - f\left(X_j^M\right) \right) \right| \le$$
$$\lim_{n\to\infty} \frac{1}{n+1} \sum_{j=0}^n \left| f\left(\tilde{X}_j^M\right) - f\left(X_j^M\right) \right|.$$

By Corollary 4.33, we have:

$$d\left(\tilde{X}_j^M, X_j^M\right) \le \frac{1}{1-s} \frac{1}{M\sqrt{2}},$$

where the bound is independent of j. Thus, since f is uniformly continuous and the right-hand side of the above inequality goes to zero as M goes to infinity, for any $\epsilon > 0$, there exists an integer $M_0 = M_0(\epsilon)$ such that, for any $M \ge M_0$ and all j, $|f(\tilde{X}_j^M) - f(X_j^M)| < \epsilon$. Hence, for all $M \ge M_0$,

$$\lim_{n\to\infty} \frac{1}{n+1} \sum_{j=0}^n \left| f\left(\tilde{X}_j^M\right) - f\left(X_j^M\right) \right| < \lim_{n\to\infty} \frac{n+1}{n+1}\epsilon = \epsilon.$$

We have thus shown that, for any $\epsilon > 0$, there exists an integer M_0 such that:

$$\left| \sum_{\tilde{x}\in\mathcal{R}_{k(M)}} f(\tilde{x})\, \pi_{k(M)}(\tilde{x}) - \int_{\mathbb{R}^2} f(x)\, d\nu(x) \right| < \epsilon, \qquad \text{for all } M \ge M_0.$$

This completes the proof.

It is known that, in general, weak convergence of a sequence of probability measures to a limiting probability measure does not imply convergence of the corresponding supports in the Hausdorff metric. However, in the case of the sequence of probability measures considered in Theorem 4.38, it is indeed true that the limit of the supports of the probability measures that comprise the sequence exists and coincides with the support of the limiting distribution. In order to prove this result, which strengthens Theorem 4.38, we shall make use of the following two lemmas.

Lemma 4.39. *Assume that (A3) is satisfied, let $\pi_{k(M)}$ be one of the $K(M)$ stationary distributions for the round-off process $\{\tilde{X}_n\}_{n=0}^{\infty}$ of accuracy $1/M$, and let $\mathcal{R}_{k(M)}$ denote its support. Let $\tilde{x} \in \mathcal{R}_{k(M)}$ be given, and let A be the support of the unique stationary distribution ν for the true process $\{X_n\}_{n=0}^{\infty}$. Then, there exists an $x \in A$ such that $d(x, \tilde{x}) \leq \theta/(1-s)$, where $\theta = 1/(M\sqrt{2})$.*

Proof. Let $B = \overline{B\left(\tilde{x}, \theta/(1-s)\right)}$ denote the closed ball of center $\tilde{x}$ and radius $\theta/(1-s)$ (regarded as a subset of $\mathbb{R}^2$), and assume that the statement of the lemma is not true, so that $A \cap B$ is empty. Since ν is supported on A, we must have $\nu(A) = 1$, which then implies that $\nu(B) = 0$. By Theorem 6.2 on page 220 in Doob (1953), letting $\tilde{X}_0 = \tilde{x}$, we have:

$$\lim_{n \to \infty} \frac{1}{n+1} \sum_{j=0}^{n} I_{\{\tilde{x}\}}\left(\tilde{X}_j\right) = \pi_{k(M)}\left(\tilde{x}\right) \qquad \text{a.s.,}$$

where $I_{\{\tilde{x}\}}(\cdot)$ denotes the indicator of the set $\{\tilde{x}\}$. On the other hand, by Elton's theorem (Theorem 3.3 on page 58) and since the ν-measure of the boundary of B is zero, letting $X_0 = \tilde{x}$, we must also have:

$$\lim_{n \to \infty} \frac{1}{n+1} \sum_{j=0}^{n} I_B\left(X_j\right) = \nu\left(B\right) \qquad \text{a.s.,}$$

where $I_B(\cdot)$ denotes the indicator of the set B.

We can therefore choose two corresponding (in the sense of Lemma 4.32) orbits $\{\tilde{x}_n\}_{n=0}^{\infty}$ and $\{x_n\}_{n=0}^{\infty}$ of $\{\tilde{X}_n\}_{n=0}^{\infty}$ and $\{X_n\}_{n=0}^{\infty}$, respectively, both starting at $\tilde{x}$, such that:

$$\lim_{n \to \infty} \frac{1}{n+1} \sum_{j=0}^{n} I_{\{\tilde{x}\}}\left(\tilde{x}_j\right) = \pi_{k(M)}\left(\tilde{x}\right) > 0$$

and

$$\lim_{n \to \infty} \frac{1}{n+1} \sum_{j=0}^{n} I_B\left(x_j\right) = \nu\left(B\right) = 0.$$

It follows from Corollary 4.33 that, whenever $\tilde{x}_j = \tilde{x}$, x_j must belong to B, so that $I_B(x_j) \geq I_{\{\tilde{x}\}}(\tilde{x}_j)$, for $j = 0, 1, 2, \ldots$. Thus, we obtain the following contradictory relationship:

$$0 = \nu(B) = \lim_{n\to\infty} \frac{1}{n+1} \sum_{j=0}^{n} I_B(x_j) \geq$$

$$\lim_{n\to\infty} \frac{1}{n+1} \sum_{j=0}^{n} I_{\{\tilde{x}\}}(\tilde{x}_j) = \pi_{k(M)}(\tilde{x}) > 0.$$

We can therefore conclude that $A \cap B$ cannot be empty, which completes the proof.

Lemma 4.40. *Under the same assumptions as in Lemma 4.39, let $x \in A$ be given. Then, there exists an $\tilde{x} \in \mathcal{R}_{k(M)}$ such that $d(\tilde{x}, x) \leq \theta/(1-s)$.*

Proof. Let $\epsilon > 0$ be given. It follows from Theorem 3.3 and a well-known consequence of weak convergence that, for any initial condition, the true chain $\{X_n\}_{n=0}^{\infty}$ will visit the closed ball $\overline{B(x, \epsilon)}$ of center x and radius ϵ infinitely often with probability one. In particular, if $\tilde{x}_0$ is any given pixel in $\mathcal{R}_{k(M)}$, there exists an orbit $\{x_n\}_{n=0}^{\infty}$ of $\{X_n\}_{n=0}^{\infty}$, with $x_0 = \tilde{x}_0$, that visits $\overline{B(x, \epsilon)}$ infinitely often. Let $\{\tilde{x}_n\}_{n=0}^{\infty}$ be the orbit of $\{\tilde{X}_n\}_{n=0}^{\infty}$ that corresponds (in the sense of Lemma 4.32) to $\{x_n\}_{n=0}^{\infty}$, and let m denote the time at which $\{x_n\}_{n=0}^{\infty}$ first enters $\overline{B(x, \epsilon)}$. Applying Corollary 4.33 and the triangle inequality, we then have:

$$d(\tilde{x}_m, x) \leq d(\tilde{x}_m, x_m) + d(x_m, x) \leq \frac{\theta}{1-s} + \epsilon.$$

Since the orbit $\{\tilde{x}_n\}_{n=0}^{\infty}$ starts at a pixel in $\mathcal{R}_{k(M)}$, it must be entirely contained in $\mathcal{R}_{k(M)}$. Hence, $\tilde{x}_m$ belongs to $\mathcal{R}_{k(M)}$. We have thus shown that, for any $\epsilon > 0$, there exists an $\tilde{x} = \tilde{x}(\epsilon) \in \mathcal{R}_{k(M)}$ such that:

$$d(\tilde{x}, x) \leq \frac{\theta}{1-s} + \epsilon. \tag{4.6}$$

Suppose now that, for all $\tilde{y} \in \mathcal{R}_{k(M)}$, we have $d(\tilde{y}, x) > \theta/(1-s)$, and let $D = \min\{d(\tilde{y}, x) \mid \tilde{y} \in \mathcal{R}_{k(M)}\}$ (since $\mathcal{R}_{k(M)}$ has finite cardinality, such a minimum must indeed be attained). D must be strictly greater than $\theta/(1-s)$ and, if we let $\epsilon = (D - \theta/(1-s))/2$, there cannot be any $\tilde{x} \in \mathcal{R}_{k(M)}$ that satisfies inequality (4.6). This contradiction proves that there must exist an $\tilde{x} \in \mathcal{R}_{k(M)}$ such that $d(\tilde{x}, x) \leq \theta/(1-s)$.

We are now in a position to state and prove the following theorem.

Theorem 4.41. *Suppose that (A3) and (A4) are satisfied. Consider the sequence of probability measures $\{\pi_{k(M)}\}_{M=1}^{\infty}$ introduced in Theorem 4.38, and let $\{\mathcal{R}_{k(M)}\}_{M=1}^{\infty}$ be the sequence of their supports. Let A be the support of ν, the unique invariant distribution for the true process $\{X_n\}_{n=0}^{\infty}$. Then, $\lim_{M \to \infty} \mathcal{R}_{k(M)} = A$, where the limit is taken with respect to the Hausdorff metric on $\mathcal{H}(\mathbb{R}^2)$, the space of nonempty, compact subsets of $\mathbb{R}^2$.*

Proof. We have previously seen that A is nonempty and compact, and that $\mathcal{R}_{k(M)}$ is nonempty, for every $M = 1, 2, \ldots$. Since, from a metric and topological point of view, we identify each pixel with its center, $\mathcal{R}_{k(M)}$ consists of the union of a finite number of points and is therefore compact. Applying Lemmas 4.39 and 4.40, and the definition of the Hausdorff metric given on page 11, we have that:

$$h\left(\mathcal{R}_{k(M)}, A\right) \leq \frac{1}{(1-s)M\sqrt{2}}, \qquad \text{for any } M = 1, 2, \ldots . \quad (4.7)$$

Since the right-hand side of the above inequality goes to zero as M goes to infinity, the statement of the theorem follows.

Figures 4.23 – 4.25 and Color Plate 7 illustrate the convergence results stated in Theorems 4.38 and 4.41 for the case of the maple leaf of Figure 2.4. Notice that a translation has been applied to the original image so as to satisfy assumption (A4), and that the viewing window has been modified accordingly.

The results presented in the aforementioned theorems have implications both on the use of the image generation algorithm based on the round-off process and on the more basic question of whether rounding errors significantly affect the accuracy with which orbits of the original process are computed in floating-point arithmetic.

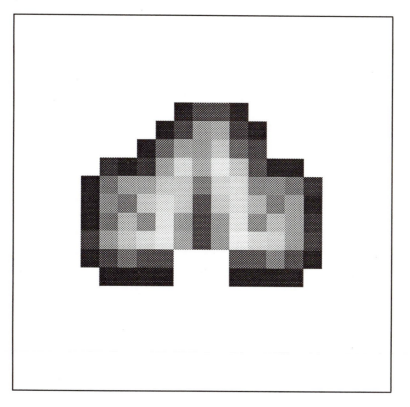

Figure 4.23. Discretized maple leaf ($M = 16$).

Specifically, as M goes to infinity, the $K(M)$ stationary distributions for the round-off process $\{\tilde{X}_n\}_{n=0}^{\infty}$ get closer (in the sense of weak convergence) to one another and to the unique stationary distribution for the true process $\{X_n\}_{n=0}^{\infty}$. This implies that, if the algorithm based on the round-off process of accuracy $1/M$ is used for the purpose of image generation, and if M is sufficiently large, the $K(M)$ distinct approximating images that can potentially be obtained do not differ significantly from one another, and they are all fairly accurate renderings of the unique image that is generated by the algorithm based on the original process. Inequality (4.7) provides an explicit upper bound on the Hausdorff distance between any of the approximating images and the true image. It is clear from the expression for the right-hand side of that inequality that the value of M required to obtain a prespecified value for the upper bound is

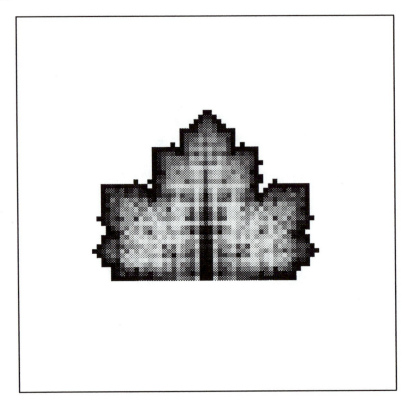

Figure 4.24. Discretized maple leaf ($M = 64$).

directly proportional to how close s is to one.

Obviously, these considerations can also be applied to the question of determining the impact of rounding errors on the evolution of orbits of the original process, with regard to the problem of assessing the closeness of the computed image to the true image. It follows from what we observed on page 72 that, if the image under consideration is contained in $[-1, 1] \times [-1, 1]$ (assumption that can always be met by applying a coordinate change), a worst-case modeling of the propagation of rounding errors can be carried out by assuming that M is equal to 2^{24}, provided that, at each step, the evaluation of $\tilde{X}_n$ can be regarded as an elementary floating-point operation (a fairly realistic assumption, for instance, when the transformations comprising the IFS are affine). Hence, on the basis of inequality (4.7), we can conclude that, if s is not extremely close to one, no appre-

Figure 4.25. Discretized maple leaf ($M = 256$).

ciable differences will be present between the shape of the computed image and the shape of the true one.

We turn now to the problem of determining under what circumstances the round-off process $\{\tilde{X}_n\}_{n=0}^{\infty}$ of accuracy $1/M$ possesses a unique stationary distribution. In order to justify the search for sufficient conditions for uniqueness, we first present some examples of IFSs whose associated round-off processes have more than one stationary distribution.

Consider the case of an affine contractive transformation w satisfying (A1) and (A2), whose round-off version $\tilde{w}$ has more than one associated minimal absorbing component. (Recall that, as a consequence of Lemma 4.31, if w is affine with fixed point at the origin, the number of minimal absorbing components is actually independent of the level of accuracy $1/M$.) The IFS $\{\mathbb{R}^2; w; p = 1\}$ gives rise to a degenerate Markov process $\{X_n\}_{n=0}^{\infty}$ and a degenerate round-off process $\{\tilde{X}_n\}_{n=0}^{\infty}$.

$\{X_n\}_{n=0}^{\infty}$ is degenerate, since, given any point $x \in \mathbb{R}^2$ and any Borel subset B of $\mathbb{R}^2$, the transition probability from x into B will either be one or zero, depending on whether $w(x)$ belongs to B or not. The unique stationary distribution for $\{X_n\}_{n=0}^{\infty}$ consists of a unit point mass at the fixed point $O = (0,0)^T$ of w.

$\{\tilde{X}_n\}_{n=0}^{\infty}$, in turn, is degenerate, since, given any pixel $\tilde{x} \in \mathcal{P}_M$, the transition probability from $\tilde{x}$ to any other pixel $\tilde{y}$ will be either one or zero, depending on whether $\tilde{y}$ equals $\tilde{w}(\tilde{x})$ or not. It is easily seen that the minimal absorbing set for $\tilde{w}$ coincides with the set of recurrent states for $\{\tilde{X}_n\}_{n=0}^{\infty}$, and that each minimal absorbing component coincides with one of the irreducible closed sets of communicating states for $\{\tilde{X}_n\}_{n=0}^{\infty}$. Thus, each minimal absorbing component is the support of a stationary distribution for $\{\tilde{X}_n\}_{n=0}^{\infty}$, and it is not difficult to verify that each such distribution is discrete uniform, due to the periodicity of the pixels in the component.

This example provides an insightful illustration of Theorems 4.38 and 4.41. For a given level of accuracy $1/M$, $\{\tilde{X}_n^M\}_{n=0}^{\infty}$ has a finite number of uniform stationary distributions supported on the minimal absorbing components for $\tilde{w}$. As M increases, the minimal absorbing components for $\tilde{w}$ tend to get closer, in the Hausdorff metric, to the fixed point O of w, and all sequences of stationary distributions for $\{\tilde{X}_n^M\}_{n=0}^{\infty}$ converge weakly to a unit point mass at O, the unique stationary distribution for $\{X_n\}_{n=0}^{\infty}$.

The next example deals with a slightly more complicated case. Consider the following two contractive transformations of the Euclidean plane into itself:

$$ w_1\left((x_1, x_2)^T\right) = \begin{pmatrix} 0.5 & 0 \\ 0 & 0.6 \end{pmatrix} \begin{pmatrix} x_1 \\ x_2 \end{pmatrix}, $$

and

$$ w_2\left((x_1, x_2)^T\right) = \begin{pmatrix} 0.5 & 0 \\ 0 & 0.6 \end{pmatrix} \begin{pmatrix} x_1 \\ x_2 \end{pmatrix} + \begin{pmatrix} 0.5 \\ 0 \end{pmatrix}. $$

The fixed point of w_1 coincides with $O = (0,0)^T$ and the fixed point of w_2 is given by $(1,0)^T$. Let $p_1 = p_2 = 1/2$. Then, the attractor of the IFS with probabilities $\{\mathbb{R}^2; w_1, w_2; p_1, p_2\}$ is given by:

$$A = \left\{ (x_1, x_2)^T \mid 0 \leq x_1 \leq 1, x_2 = 0 \right\},$$

and the unique stationary distribution for the associated Markov chain $\{X_n\}_{n=0}^{\infty}$ is uniform on A.

Consider now the round-off process $\{\tilde{X}_n\}_{n=0}^{\infty}$ of accuracy $1/M$ associated with the given IFS. The results that we are going to outline without formal proof can be easily verified by means of arguments similar to those employed in Section 4.3 to describe the behavior of orbits under the round-off versions of contractive transformations of the Euclidean plane into itself. We have chosen to omit the details of the analysis because they might hinder comprehension and detract from the essence of the example. For the purpose of illustration, we shall restrict our attention to the case $M = 5$, extension to the general case being similar and straightforward.

We observe first that every pixel $P(i, j)$ whose j coordinate is strictly greater than one in absolute value, or whose i coordinate is outside of the 1 to 5 range, is transient for $\{\tilde{X}_n^5\}_{n=0}^{\infty}$. All other pixels are recurrent and can be partitioned into three irreducible closed sets of communicating states given by:

$$\mathcal{R}_k = \{P(i, j) \in \mathcal{P}_5 \mid 1 \leq i \leq 5, j = k\}, \quad \text{for } k = -1, 0, 1.$$

Observe that this partition is brought about by the fact that a contractivity factor of 0.6 along the vertical axis is too large for pixels in $\mathcal{R}_{-1}$ and $\mathcal{R}_1$ to be able to communicate with pixels in $\mathcal{R}_0$.

If we denote a pixel $P(i, k)$ in $\mathcal{R}_k$ by its i coordinate, it is easily verified that the transition probability matrix for the Markov chain $\{\tilde{X}_n^5\}_{n=0}^{\infty}$ restricted to $\mathcal{R}_k$ is given by:

	1	2	3	4	5
1	1/2	0	1/2	0	0
2	1/2	0	0	1/2	0
3	0	1/2	0	1/2	0
4	0	1/2	0	0	1/2
5	0	0	1/2	0	1/2

$, \quad$ for $k = -1, 0, 1.$

Since this matrix is doubly stochastic, it follows that the unique stationary distribution for the restriction of $\{\tilde{X}_n^5\}_{n=0}^{\infty}$ to $\mathcal{R}_k$ is discrete uniform, for $k = -1, 0, 1.$

As previously mentioned, there is nothing special about the case $M = 5$, and, for a generic level of accuracy $1/M$, the round-off process $\{\tilde{X}_n^M\}_{n=0}^{\infty}$ will have as its stationary distributions π_k the three discrete uniform distributions supported on:

$$\mathcal{R}_k = \{P(i,j) \in \mathcal{P}_M \,|\, 1 \le i \le M, j = k\}, \quad \text{for } k = -1, 0, 1.$$

Furthermore, the empirical distributions of almost every orbit of $\{\tilde{X}_n^M\}_{n=0}^{\infty}$ starting at a pixel $P(i,j)$, with $j \ge 1$, will converge weakly to π_1; the empirical distributions of almost every orbit starting at a pixel $P(i,j)$, with $j = 0$, will converge weakly to π_0; and the empirical distributions of almost every orbit starting at a pixel $P(i,j)$, with $j \le -1$, will converge weakly to π_{-1}.

It should be clear from this discussion how Theorems 4.38 and 4.41 apply to the present example. Any sequence of distributions constructed by choosing at each stage M one of the three discrete uniform stationary distributions for $\{\tilde{X}_n^M\}_{n=0}^{\infty}$ converges weakly, as M goes to infinity, to the continuous uniform distribution on A, which is the unique stationary distribution for $\{X_n\}_{n=0}^{\infty}$, while the supports of the distributions in the sequence converge to A in the Hausdorff metric. In particular, for the practical purpose of image generation, if M is sufficiently large, any of the three discrete uniform distributions will provide a satisfactory approximation to the continuous uniform distribution on A.

Having presented examples of IFSs whose associated round-off processes of given accuracy $1/M$ possess more than one stationary distribution, we shall now state two sufficient conditions for uniqueness. The first condition is given in the following theorem, whose proof is a direct consequence of Lemma 4.36.

Theorem 4.42. *Assume that (A3) and (A4) are satisfied. If $\tilde{w}_1$, the round-off version of w_1 of accuracy $1/M$, has only one minimal absorbing component, then the round-off process $\{\tilde{X}_n\}_{n=0}^{\infty}$ of accuracy $1/M$ has a unique stationary distribution.*

Lemmas 4.25 and 4.26 imply the following corollary.

Corollary 4.43. *Under assumptions (A3) and (A4), the round-off process $\{\tilde{X}_n\}_{n=0}^{\infty}$ of accuracy $1/M$ has a unique stationary distri-*

bution if w_1 has a contractivity factor s_1 which is strictly less than $1/2$.

Remark 4.44. The previous theorem and corollary, as well as the following theorem, provide conditions for uniqueness based on properties of w_1. It is clear that a relabeling of the transformations in the IFS and a translation of the coordinate system is all that is needed to meet the assumptions, provided that at least one of the transformations in the IFS satisfies the required properties. In particular, when applying Corollary 4.43, it might be convenient to relabel the transformations and choose the coordinate system in such a way that assumption (A5) is satisfied.

The condition for uniqueness given in the next theorem is based on the existence of a certain relationship between the attractor of the IFS and the domains of attraction of the round-off version of one of the transformations in the IFS.

Theorem 4.45. *Assume that (A3) and (A4) are satisfied, and let A denote the attractor of the IFS. If there exists an $x \in A$ and an $\epsilon > 0$ such that the set*

$$\mathcal{B}(x, \epsilon) = \left\{ \tilde{x} \in \mathcal{P}_M \,\middle|\, d(\tilde{x}, x) \leq \frac{\theta}{1 - s} + \epsilon \right\}$$

is contained in the domain of attraction $\mathcal{D}$ of one of the minimal absorbing components for $\tilde{w}_1$, then the round-off process $\{\tilde{X}_n\}_{n=0}^{\infty}$ of accuracy $1/M$ has a unique stationary distribution. Here, as usual, $\theta = 1/(M\sqrt{2})$ denotes half the length of a pixel diagonal.

Proof. Let $\mathcal{C}$ be the minimal absorbing component for $\tilde{w}_1$ whose domain of attraction is given by $\mathcal{D}$. The argument given at the beginning of the proof of Lemma 4.36 implies that we only need to show that any pixel $\tilde{x}$ which is recurrent for $\{\tilde{X}_n\}_{n=0}^{\infty}$ communicates with a pixel in $\mathcal{C}$. By Elton's theorem (Theorem 3.3 on page 58) and a well-known consequence of weak convergence, the original chain starting at $\tilde{x}$ will visit the closed ball $\overline{B(x, \epsilon)}$ of center x and radius ϵ infinitely often with probability one. This implies that there exists a trajectory $\{x_{n+1} = w_{\sigma_n}(x_n)\}_{n=0}^{\infty}$ of $\{X_n\}_{n=0}^{\infty}$ starting at $\tilde{x}$, and an integer $m_1 \geq 0$, such that $x_{m_1} \in \overline{B(x, \epsilon)}$. Corollary 4.33 and the triangle inequality imply that, for the corresponding orbit of $\{\tilde{X}_n\}_{n=0}^{\infty}$, we must have:

$$d(\tilde{x}_{m_1}, x) \leq d(\tilde{x}_{m_1}, x_{m_1}) + d(x_{m_1}, x) \leq \frac{\theta}{1-s} + \epsilon,$$

so that $\tilde{x}_{m_1} \in B(x, \epsilon)$. Since $B(x, \epsilon)$ is contained in $\mathcal{D}$, there exists an integer $m_2 \geq 0$ such that the orbit of $\tilde{x}_{m_1}$ under $\tilde{w}_1$ will first enter $\mathcal{C}$ at time m_2. Hence, there exists a path of $\{\tilde{X}_n\}_{n=0}^{\infty}$ that leads from $\tilde{x}$ to $\tilde{w}_1^{\circ m_2}(\tilde{x}_{m_1}) \in \mathcal{C}$ with positive probability $p = p_{\sigma_0} \cdots p_{\sigma_{m_1-1}} p_1^{m_2}$. If there were no path leading back to $\tilde{x}$, the chain $\{\tilde{X}_n\}_{n=0}^{\infty}$ starting at $\tilde{x}$ would then have a positive probability of at least p of never returning to $\tilde{x}$, contradicting the hypothesis that $\tilde{x}$ is recurrent. The argument given at the end of the proof of Lemma 4.36 can now be repeated to conclude that all recurrent states for $\{\tilde{X}_n\}_{n=0}^{\infty}$ belong to the same irreducible closed set. This, in turn, implies uniqueness of the stationary distribution for $\{\tilde{X}_n\}_{n=0}^{\infty}$.

In the example starting on page 138, neither the conditions of Theorem 4.42 nor those of Theorem 4.45 are satisfied. As far as Theorem 4.42 is concerned, it is easily verified that, for any level of accuracy $1/M$, the minimal absorbing set for $\tilde{w}_1$ is given by $\mathcal{M} = \{P(i, j) \in \mathcal{P}_M | 0 \leq i \leq 1, -1 \leq j \leq 1\}$, with each of the six pixels in $\mathcal{M}$ being an individual minimal absorbing component. In order to verify that the conditions of Theorem 4.45 are not satisfied, observe first that the smallest contractivity factor for the IFS is $s = 0.6$. Now, let $x \in A = [0, 1] \times \{0\}$ be given and, for a given level of accuracy $1/M$, let $P(i, 0)$, with $0 \leq i \leq M$, be the pixel to which x belongs. We then have:

$$d\left(P(i, 0), x\right) \leq \frac{1}{2M} < \frac{5}{2M\sqrt{2}} = \frac{\theta}{1-s}.$$

Furthermore:

$$d\left(P(i, 1), x\right) \leq d\left(P(i, 1), P(i, 0)\right) + d\left(P(i, 0), x\right) \leq$$
$$\frac{1}{M} + \frac{1}{2M} = \frac{3}{2M} < \frac{5}{2M\sqrt{2}} = \frac{\theta}{1-s}.$$

It follows that, for any $\epsilon > 0$, both $P(i, 0)$ and $P(i, 1)$ must belong to $\mathcal{B}(x, \epsilon)$. However, it is immediately verified that these two pixels belong to the domains of attraction of two distinct minimal absorbing components for $\tilde{w}_1$, so that the assumptions of the theorem are not met.

We shall now give an example that illustrates how Theorem 4.45 can be applied in practice. Consider the IFS with probabilities $\{\mathbb{R}^2; w_1, w_2, w_3; p_1 = p_2 = p_3 = 1/3\}$, with

$$w_1\left((x_1, x_2)^T\right) = \begin{pmatrix} 0.6 & 0 \\ 0 & 0.6 \end{pmatrix}\begin{pmatrix} x_1 \\ x_2 \end{pmatrix},$$

$$w_2\left((x_1, x_2)^T\right) = \begin{pmatrix} 0.6 & 0 \\ 0 & 0.6 \end{pmatrix}\begin{pmatrix} x_1 \\ x_2 \end{pmatrix} + \begin{pmatrix} 0.4 \\ 0 \end{pmatrix},$$

$$w_3\left((x_1, x_2)^T\right) = \begin{pmatrix} 0.6 & 0 \\ 0 & 0.6 \end{pmatrix}\begin{pmatrix} x_1 \\ x_2 \end{pmatrix} + \begin{pmatrix} 0.2 \\ 0.4 \end{pmatrix},$$

whose attractor is illustrated in Figure 4.26.

Although assumptions (A3) and (A4) are satisfied, the round-off version of w_1 has, for any level of accuracy $1/M$, nine distinct minimal absorbing components (see Section 4.3), and Theorem 4.42 cannot be applied to conclude that the round-off process $\{\tilde{X}_n\}_{n=0}^\infty$ has a unique stationary distribution. However, uniqueness of the stationary distribution for $\{\tilde{X}_n\}_{n=0}^\infty$ can be guaranteed, for any level of accuracy $1/M$, with $M \geq 4$, by means of Theorem 4.45.

To this end, observe first that the point $x = (1/2, 1)^T$, being the fixed point of w_3, must belong to the attractor A by Lemma 2.3. It is also easily verified that the IFS has a contractivity factor $s = 0.6$. From the analysis presented in Section 4.3, we know that, for any level of accuracy $1/M$, pixel $P(1, 1)$ constitutes a minimal absorbing component for $\tilde{w}_1$, whose domain of attraction is given by:

$$\mathcal{D} = \{P(i, j) \in \mathcal{P}_M | i \geq 1, j \geq 1\}.$$

It is clear that, if a pixel does not belong to $\mathcal{D}$, its distance from $x = (1/2, 1)^T$ must be at least $1/2$. It follows that the sufficient condition of Theorem 4.45 is certainly satisfied if $\theta/(1-s) + \epsilon < 1/2$, for some $\epsilon > 0$. Substituting $\theta = 1/(M\sqrt{2})$ and $s = 6/10$ into the previous inequality yields $M > 5/\sqrt{2} + \epsilon$. Since $3 < 5/\sqrt{2} < 4$,

Figure 4.26. Attractor for the IFS in the last example of the chapter.

the previous condition can be satisfied for any integer $M \geq 4$ if we take, for instance, $\epsilon = (4 - 5/\sqrt{2})/2$. Thus, the round-off process $\{\tilde{X}_n\}_{n=0}^{\infty}$ of accuracy $1/M$ associated with the given IFS has a unique stationary distribution for any $M \geq 4$.

We would like to conclude this chapter with some brief remarks on a new, possible way of addressing the image encoding problem, based on the properties of the round-off process. Suppose that a given color image is known to have been generated through the random iteration algorithm based on the round-off process $\{\tilde{X}_n\}_{n=0}^{\infty}$ of accuracy $1/M$ associated with an unknown IFS with probabilities $\{\mathbb{R}^2; w_1, \ldots, w_N; p_1, \ldots, p_N\}$. The level of accuracy $1/M$ can be inferred from the image, and the number N of transformations comprising the IFS is assumed, for simplicity, to be known. In addition, all the transformations in the IFS are assumed to be affine.

By virtue of the familiar correspondence between frequencies and colors, having the image at our disposal is tantamount to knowing one of the stationary distributions for the round-off process. Assume that the recorded frequencies for the pixels belonging to the support $\mathcal{R}$ of the stationary distribution under consideration have been arranged in a row vector π (with entries $\pi(\tilde{x})$, $\tilde{x} \in \mathcal{R}$), and let P denote the transition probability matrix for the Markov chain $\{\tilde{X}_n\}_{n=0}^{\infty}$ restricted to $\mathcal{R}$, where the ordering of the entries is consistent with the arrangement of the entries of π. The matrix P will have entries:

$$
\begin{aligned}
P(\tilde{x}, \tilde{y}) &= P\left\{\tilde{X}_{n+1} = \tilde{y} \,\middle|\, \tilde{X}_n = \tilde{x}\right\} \\
&= \sum_{i=1}^{N} p_i I_{\{\tilde{y}\}}\left(\tilde{w}_i(\tilde{x})\right), \qquad \text{for } \tilde{x}, \tilde{y} \in \mathcal{R}, \qquad (4.8)
\end{aligned}
$$

where $I_{\{\tilde{y}\}}(\cdot)$ denotes the indicator of the set $\{\tilde{y}\}$.

Since π is the unique stationary distribution for the restriction of $\{\tilde{X}_n\}_{n=0}^{\infty}$ to $\mathcal{R}$, it must be the unique vector with positive entries adding up to one which satisfies the invariance condition $\pi = \pi P$ (see Ross (1983) for detail). It is clear from Equation (4.8) that the entries of the matrix P are functions of the unknown parameters of the transformations comprising the IFS and of the associated probabilities. (Notice, incidentally, that the only nonzero entries in a given row of P should occur in the locations which correspond to affine images of the state $\tilde{x}$ corresponding to that row.) The encoding problem can then be solved, in principle, by minimizing the Euclidean norm of $(\pi - \pi P)$ as a function of the unknown parameters and probabilities, subject to the constraint that the resulting IFS is hyperbolic.

Considerations analogous to those presented in Section 2.5 apply to this case. In particular, our preliminary, limited experience is that, in general, minimization of the proposed objective function is a difficult numerical problem, due to the aforementioned constraint and to the presence of several local minima. On the brighter side, from a computational point of view, function evaluations are much simpler in the present case than when determination of the Hausdorff distance between subsets of $\mathbb{R}^2$ is required, in order to minimize the left-hand side of Equation (2.4).

5

Animation

The method of image generation introduced in the second and third chapters can also be successfully employed to produce computer animated motion pictures. As is well-known, a traditional motion picture consists of a series of still pictures, called frames, projected on a screen in rapid succession. The objects in the frames are shown in successive positions slightly changed, so as to produce the optical effect of a continuous picture in which the objects move. Nowadays, motion pictures can be conveniently recorded on videotapes for playing through a television set. Such recordings are commonly referred to as videos. In the United States, frames are shown on television at a rate of 30 per second. At that speed, if the frames vary smoothly, the human eye is incapable of seeing the individual still pictures and perceives motion.

The random iteration algorithm based on IFS theory can be employed to generate on a computer the individual frames of an animation sequence. Such frames can then be successively displayed on a monitor or recorded on a videotape. (The technical details of the recording procedure will be described in Section 5.4. We also refer the reader to Foley et al. (1990) for an excellent and comprehensive treatment of all aspects of computer graphics, including computer animation.) The animation procedure hinges on the fact that, when all transformations in a continuously parameterized family of IFSs are strictly contractive, the corresponding images vary continuously (in the Hausdorff metric) with the parameter indexing the family.

For instance, in the case when all transformations are affine, interpolation between two images can be attained by simple interpolation between the coefficients of the mappings. More specifically, suppose it is desired to produce an animation segment whose first frame is image A and whose last frame is image B, and assume, for the sake of simplicity, that both images can be generated through hyperbolic IFSs consisting of the same number of affine transformations. Then, if one pairs each transformation that generates A to one of the transformations that generate B, and constructs new sets of transformations by linearly interpolating between the coefficients of every pair, the images corresponding to the new IFSs can be used as intermediate frames of the animation segment. In the probabilistic case, the values of the probabilities associated with each mapping can also be determined through linear interpolation. Notice that other intermediate frames may be obtained by interpolating the transformations and associated probabilities in a nonlinear fashion.

Many different movies, starting with the same frame A and ending with the same frame B, can be produced by simply varying the way in which the transformations are initially paired. More precisely, if both A and B can be generated with N transformations, the total number of such pairings equals $N!$. From a computational point of view, animation is an application where parallelism is really efficient, since all frames can be generated simultaneously.

The first two sections in this chapter deal with the statement of the continuity results upon which the animation procedure is based, and with an example in which continuity fails to hold, because not all assumptions are met. The third section contains examples and suggestions about practical ways in which the ideas developed in the two preceding sections can be implemented. Finally, the last section in the chapter presents a description of the animation equipment that we use to record computer-animated motion pictures on videotape.

5.1. Continuity Considerations

In this section, we shall present the basic technical results needed to apply IFS theory to the production of computer-animated motion pictures. As noted in the introduction, what is necessary for the animation technique to work is some sort of continuous dependence of the attractors of IFSs upon some suitable parameterization of the transformations. The two lemmas and the theorem that follow, stated and proved in Barnsley (1988), provide useful results in that direction.

Lemma 5.1. *Assume that (Δ, d_δ) is a metric space and that (X, d) is a complete metric space. Suppose that $w : \Delta \times X \mapsto X$ is such that there exists an s, with $0 \le s < 1$, such that, for any $\delta \in \Delta$, the mapping $x \mapsto w(\delta, x)$ is a contraction, with contractivity factor s. Assume also that, for any $x \in X$, the mapping $\delta \mapsto w(\delta, x)$ is continuous. For each $\delta \in \Delta$, let $x_f(\delta)$ denote the fixed point of $w(\delta, \cdot)$. Then $x_f : \Delta \mapsto X$ is continuous.*

Lemma 5.2. *Let (Δ, d_δ) be a compact metric space, and let (X, d) be a metric space. For $n = 1, \ldots, N$, let $w_n : \Delta \times X \mapsto X$ be jointly continuous. Define $W : \Delta \times \mathcal{H}(X) \mapsto \mathcal{H}(X)$ by:*

$$W(\delta, B) = \bigcup_{n=1}^{N} w_n(\delta, B), \quad \text{for any } \delta \in \Delta, \text{ and } B \in \mathcal{H}(X),$$

where $\mathcal{H}(X)$ denotes, as usual, the collection of nonempty, compact subsets of X, endowed with the Hausdorff metric. Then, for every $B \in \mathcal{H}(X)$, the mapping $\delta \mapsto W(\delta, B)$ is continuous.

Combining Lemmas 5.1 and 5.2 we obtain the following fundamental theorem.

Theorem 5.3. *Suppose that the conditions of Lemma 5.2 are satisfied, and assume also that (X, d) is complete. Suppose there exists an s, with $0 \le s < 1$, such that, for any $\delta \in \Delta$ and for $n = 1, \ldots, N$, the mapping $x \mapsto w_n(\delta, x)$ is a contraction, with contractivity factor s. Then, for each $\delta \in \Delta$, there is an attractor $A(\delta) \in \mathcal{H}(X)$*

such that $W(\delta, A(\delta)) = A(\delta)$. Furthermore, the mapping $\delta \mapsto A(\delta)$ is continuous.

Suppose now that $\{\mathbb{R}^2; u_1, \ldots, u_N\}$ and $\{\mathbb{R}^2; v_1, \ldots, v_N\}$ are two hyperbolic IFSs on the Euclidean plane with respective contractivity factors s_u and s_v, and let $s = \max(s_u, s_v)$. Let $\Delta = [0, 1]$ be endowed with the Euclidean metric. For $n = 1, \ldots, N$, define:

$$w_n(\delta, x) = \delta u_n(x) + (1 - \delta)v_n(x), \quad \text{for any } \delta \in \Delta, \text{ and } x \in \mathbb{R}^2.$$
(5.1)

Then, for any $\delta \in \Delta$, and any $x, y \in \mathbb{R}^2$, we have:

$$\begin{aligned}
d(w_n(\delta, x), w_n(\delta, y)) &= \\
&= |(\delta u_n(y) + (1 - \delta)v_n(y)) - (\delta u_n(x) + (1 - \delta)v_n(x))| \\
&\leq \delta|u_n(y) - u_n(x)| + (1 - \delta)|v_n(y) - v_n(x)| \\
&\leq \delta s_u|y - x| + (1 - \delta)s_v|y - x| \\
&\leq \delta s\, d(x, y) + (1 - \delta)s\, d(x, y) \\
&= s\, d(x, y),
\end{aligned}$$

so that, for any $\delta \in \Delta$, and $n = 1, \ldots, N$, $w(\delta, x)$ is a contraction with contractivity factor s, $0 \leq s < 1$. Since all transformations w_n, being given by a convex combination of continuous functions, are obviously jointly continuous, and all other assumptions are satisfied, we can apply Theorem 5.3 and conclude that the function that associates with each IFS $\{\mathbb{R}^2; w_1(\delta, x), \ldots, w_N(\delta, x)\}$ its attractor $A(\delta)$ is continuous in δ.

This result can be readily utilized to produce an animation segment. We start with two hyperbolic IFSs consisting only, for the sake of computational ease, of affine transformations of the Euclidean plane into itself, and consider the family of IFSs defined by Equation (5.1). The actual construction of any IFS in such a family is computationally very simple, since, for any $\delta \in \Delta$, the parameters of the transformations $w_n(\delta, x)$, $n = 1, \ldots, N$, can be easily obtained by linear interpolation between the corresponding parameters of u_n and v_n. We then consider N_F distinct values of δ, equally spaced between 0 and 1, and generate, for each such value, the attractor $A(\delta)$ of the corresponding IFS. [1] The N_F images thus obtained will

---

[1] See Foley et al. (1990) for a discussion of why, in certain circumstances, a spacing

constitute the individual frames of an animation segment whose first
and last frame will be given by $A(0)$ and $A(1)$, the attractors of the
two original IFSs. Smoothness in the transition from $A(0)$ to $A(1)$ is
guaranteed by the continuity result stated in Theorem 5.3, provided
the number of frames is sufficiently large.

Note that the choice of determining the intermediate IFSs through
linear interpolation is doubly expedient, since, on the one hand, it
is easy to implement and, on the other, it ensures that the intermediate IFSs are hyperbolic. Consideration of IFSs whose transformations are not necessarily affine is also possible, and can be handled
similarly. The main reasons why we prefer to deal with affine transformations lie in their manageability, and in the simplicity of affine
arithmetic.

While we have, so far, restricted our attention to deterministic
IFSs, and, therefore, black-and-white images, consideration of IFSs
with probabilities and associated color images is similar and straightforward. In fact, it is possible to prove that, under conditions that
are a direct extension of the hypotheses of Theorem 5.3 to the probabilistic case, the invariant probability of a hyperbolic IFS with probabilities is a continuous function, with respect to the Hutchinson
metric, of a suitable parameterization of the transformations and
of the associated probabilities. In particular, given two hyperbolic
IFSs with probabilities consisting only of affine transformations, it
is possible to produce a full-color animation segment by linear interpolation between the parameters of corresponding transformations
and their associated probabilities. The intermediate frames will then
be given by the supports of the invariant probabilities of the inter-
polated hyperbolic IFSs, with color being added by means of the
familiar correspondence between frequencies and colors.

5.2. A Discontinuity Example

We shall now discuss the importance of the assumption made in the
previous section that, for all values of the parameter δ, the corresponding IFS is hyperbolic. The idea of considering nonhyperbolic

---

other than uniform might be preferable.

IFSs could be justified, in theory, by the fact that, in light of Theorem 3.1, not all transformations need be contractive in order for an IFS with probabilities to have a unique invariant probability. In particular, we could, as we did in the previous section, parameterize a family of IFSs with probabilities which are not necessarily hyperbolic but each of which still has a unique invariant probability, and hope for continuous dependence of the supports of the invariant probabilities on the given parameterization. The following example shows that such a continuous dependence need not hold.

Consider the IFS with probabilities $\{\mathbb{R}^2; w_1(x), w_2(x); p_1, p_2\}$ on the Euclidean plane, where

$$w_1\left((x_1, x_2)^T\right) = \begin{pmatrix} 0 & 0 \\ 0 & 0 \end{pmatrix}\begin{pmatrix} x_1 \\ x_2 \end{pmatrix} + \begin{pmatrix} 1 \\ 0 \end{pmatrix}$$

and

$$w_2\left((x_1, x_2)^T\right) = \begin{pmatrix} \cos\theta & \sin\theta \\ -\sin\theta & \cos\theta \end{pmatrix}\begin{pmatrix} x_1 \\ x_2 \end{pmatrix},$$

with $0 < \theta < \pi/2$, and θ/π irrational.

Since the norm of the matrix that appears in the definition of the latter transformation equals one, w_2 is not a strict contraction and the IFS is not hyperbolic. However, it is easily seen that, for any assignment of nonzero probabilities p_1 and p_2, the IFS has a unique invariant probability, whose support A is given by the circle of radius one centered at the origin. Notice that, in this case, A is no longer the *unique* compact set satisfying the self-covering condition (3.2), the closed unit disk centered at the origin being an example of another compact self-covering set.

For any $\delta \in [0, \cos\theta]$, consider the IFS with probabilities $\{\mathbb{R}^2; w_1(\delta, x), w_2(\delta, x); p_1, p_2\}$, where

$$w_1\left(\delta, (x_1, x_2)^T\right) = \begin{pmatrix} 0 & 0 \\ 0 & 0 \end{pmatrix}\begin{pmatrix} x_1 \\ x_2 \end{pmatrix} + \begin{pmatrix} 1 \\ 0 \end{pmatrix}$$

and

$$w_2\left(\delta, (x_1, x_2)^T\right) = \begin{pmatrix} \cos\theta - \delta & \sin\theta \\ -\sin\theta & \cos\theta - \delta \end{pmatrix}\begin{pmatrix} x_1 \\ x_2 \end{pmatrix}.$$

It is not difficult to prove that, for any $\delta \in (0, \cos\theta]$, such an IFS is hyperbolic and has, therefore, a unique invariant probability, whose support is independent of the particular nonzero values of p_1 and p_2, and coincides, when the IFS is regarded as deterministic, with its unique compact attractor $A(\delta)$. We can then define, as in Theorem 5.3, a function $A(\delta)$ which maps every $\delta \in (0, \cos\theta]$ into the attractor of the corresponding IFS. It seems natural, at this point, to extend the definition of such a function to $\delta = 0$ by setting $A(0) = A$, with A being the support of the invariant probability of the IFS with probabilities corresponding to $\delta = 0$. At this point, letting $\Delta = [0, \cos\theta]$, all the assumptions of Theorem 5.3 are satisfied, except for the fact that the IFS corresponding to $\delta = 0$ is not hyperbolic, so that the definition of the mapping $A(\delta)$ at $\delta = 0$ had to be suitably modified.

We shall now show that $A(\delta)$ is not continuous at $\delta = 0$. Observe first that the fixed point of $w_2(\delta, x)$ is given by $x_0 = (0, 0)^T$, for any $\delta \in (0, \cos\theta]$. Since, as a simple consequence of Lemma 2.3, the fixed point of any transformation in a hyperbolic IFS must belong to the attractor of the IFS, x_0 belongs to $A(\delta)$, for any $\delta \in (0, \cos\theta]$. Then, being that $d(x, x_0) = 1$ for any $x \in A(0)$, the Hausdorff distance between $A(\delta)$ and $A(0)$ must satisfy the relation $h(A(\delta), A(0)) \geq 1$, for any $\delta \in (0, \cos\theta]$, so that $A(\delta)$ cannot be continuous at $\delta = 0$. It is actually possible to show that the limit as δ goes to zero of $A(\delta)$ is the closed unit disk centered at the origin.

It is clear from the previous example that, in order to obtain smooth animation, it is important that all IFSs in the parameterized family be hyperbolic, and that, in general, it is not possible to relax such a condition by extending the definition of the mapping $A(\delta)$ in the way that has been suggested. As shown in the previous section, one way to guarantee that the family of IFSs at hand satisfies this requirement is to construct it by linear interpolation between two hyperbolic IFSs, whose corresponding images will then constitute the first and last frames of the animation segment. This does not mean that other parameterizations should not be considered. However, when using different approaches to construct the family of IFSs to be employed to produce animation, one should always make sure that all transformations involved be strictly contractive, so as to guarantee smoothness in the resulting animation segment.

5.3. Making Movies

In this section, we shall analyze in more detail issues related to the animation procedure. There are several reasons why the possibility of applying IFS techniques to the production of motion pictures is of interest to us, and we would like to briefly present a few of them.

It is indeed crucial, if IFS theory is to become practically applicable, that a good understanding be gained about the types of images that can be encoded by means of IFSs. The interpolation procedure described in Section 5.1 provides a straightforward way of constructing a large number of IFSs starting with just two of them. By generating their associated images, and exploiting the continuity property, it becomes possible, as we explained before, to produce a smooth animation segment. It seems reasonable, then, to envision the possibility of joining together several such segments and record them on videotape, thus creating a library of images that can be encoded through IFSs. Such a library could eventually become a useful source from which to extract the building blocks to be used for creating more elaborate movies.

From an even more practical point of view, this animation procedure adds yet another option to the many available to any person interested in creating an animated motion picture. In view of the self-covering property that characterizes their attractors, IFSs are extremely well-suited for encoding images that have patterns that tend to repeat themselves at any level of magnification. The maple leaf of Figure 2.4, the fern of Figure 2.5, and the pine tree of Figure 2.8, all illustrate natural objects that have such a property, and can be encoded with just a few affine transformations. In general, it appears that IFS encoding is extremely effective when applied to a large variety of natural objects, such as clouds, coastlines, trees, forests, mountain profiles, and so on. This characteristic suggests that IFS techniques can be profitably employed, for instance, to generate, among other things, landscapes and natural scenes to be used as backgrounds for animated motion pictures.

On a related note, focus has recently been placed on the possibility of employing IFS encoding in developing flight simulators. This constitutes an application in which there is great need for data compression. Typically, one would like to be able to generate and display on a monitor, in real-time, a continuous sequence of images of natural surroundings, so as to simulate the environment encountered by a pilot during flight. As we have already remarked several times, standard representation of digitized images requires an enormous amount of storage space, so that it becomes difficult to find storage media that can hold all the frames necessary to produce an animated movie of reasonable length. On the other hand, if each frame can be encoded through an IFS consisting of affine transformations, only the parameters of the transformations and the associated probabilities must be recorded, and much storage space can be saved. As noted in Chapter 3, image decoding can be attained in a speedy and efficient manner by means of the random iteration algorithm, and it is indeed conceivable that real-time generation of a movie can actually be achieved through either parallel implementation or the use of customized hardware.

We would now like to turn our discussion to slightly more technical matters related to our experience in the field of animation. As we said before, animation can be used to explore the world of images that can be generated through IFS encoding. We have done so in a way that, although far from being systematic, has allowed us, on the one hand, to discover some fascinating images, and, on the other, to increase our familiarity with the animation procedure itself.

As a first example, consider the six images of Figure 5.1. We immediately recognize that the first image coincides with the maple leaf of Figure 2.4, which can be encoded through the four affine transformations listed on page 22, and the associated probabilities listed on page 44. The last image, on the other hand, represents the familiar fern of Figure 2.5, whose encoding is attained through the four affine transformations listed on page 23, and the associated probabilities listed on page 44. The remaining four images are equally spaced intermediate frames of an animation segment that begins with the maple leaf and ends with the fern. They have been obtained according to the linear interpolation scheme described in Section 5.1, where each transformation in the encoding of the leaf has been paired with the transformation in the encoding of the fern which is labeled with the same number. Notice how, in the smooth transition from the first to the last frame, the maple leaf almost shrinks to a point and eventually unfolds back into the fern.

Figure 5.1. From leaf to fern (I).

As pointed out in the introduction to this chapter, by pairing the transformations in a different order, several animation segments can be obtained. As an example, compare Figure 5.1 to Figure 5.2, in which the intermediate frames have been generated by interpolating

Figure 5.2. From leaf to fern (II).

the first transformation in the encoding of the maple leaf with the fourth in the encoding of the fern, the second with the third, the third with the second, and the fourth with the first.

In outlining how interpolation between two images can be achieved, and in all the examples presented so far, we have always dealt with cases in which the first and last frame of the animation segment consist of images that can both be encoded through IFSs which contain the same number of transformations. If this is not the case, there are still several ways in which smooth interpolation can be attained.

It follows from the self-covering property stated in Equation (3.2) that the shape of the attractor A of an IFS does not change if we add to its encoding a contractive transformation w, such that $w(A) \subset A$. Now, for example, the maple leaf, as we have just seen, can be encoded with four affine transformations, while the dragon of Figure 2.3 can be encoded with the two transformations listed on page 21. If we repeat twice each transformation in the encoding of the dragon, and assign probability $1/4$ to each of the four transformations, we obtain a new IFS encoding for the dragon that consists of the same number of transformations used to encode the maple leaf. Figure 5.3 contains six equally spaced frames of an animation segment generated using the usual linear interpolation scheme, where the first transformation and the second transformation in the encoding of the leaf have each been paired with a copy of the first transformation in the original encoding of the dragon, while the third transformation and the fourth transformation in the encoding of the leaf have each been paired with a copy of the second transformation in the encoding of the dragon.

The example illustrated in Figure 5.4 has been constructed in a similar fashion, by adding two degenerate transformations that respectively map the entire plane into the fixed points of the first transformation and of the second transformation in the original encoding of the dragon, and assigning probability zero to each of them, while leaving the probabilities associated with the two original transformations unaltered. The pairing employed in the linear interpolation matches the first transformation and the fourth transformation in the encoding of the leaf to the first degenerate transformation and the second degenerate transformation introduced above, and matches the second transformation and the third transformation in the encoding of the leaf to the first transformation and the second transformation in the original encoding of the dragon.

An additional example of an animation segment is presented in Figure 5.5, which illustrates a possible transition from the spiral of Figure 2.7 to the twin dragon. Notice how some of the intermediate

Figure 5.3. From leaf to dragon (I).

frames present cloudlike shapes that could be used when trying to render the texture of a sky.

In all the examples presented so far, we have limited our intervention to the choice of the first and last frame of each animation

Figure 5.4. From leaf to dragon (II).

segment, and to the pairing of the transformations, without exerting any further control over the intermediate frames. One important way of controlling what the final animation segment will look like is by means of suitable coordinate changes.

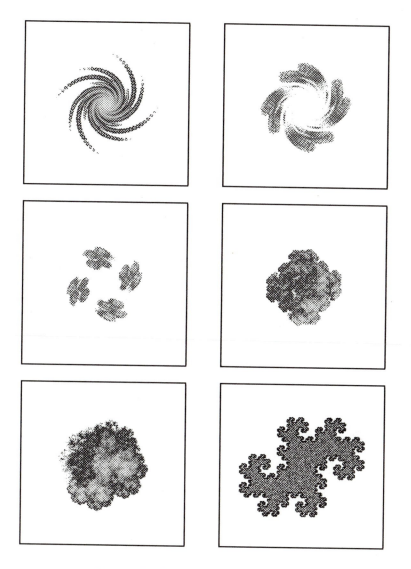

Figure 5.5. From spiral galaxy to dragon.

A coordinate change on a metric space X is an invertible transformation θ of X into itself. If x are the coordinates of a point in the original coordinate system, $x' = \theta(x)$ will denote its coordinates in the new system. Suppose now that w is a given transformation of

X into itself, whose formula is expressed in the original coordinate system. It is shown in Barnsley (1988) that, if w' denotes the formula for the same transformation expressed in the new coordinate system, the following relationships hold:

$$w(x) = \left(\theta^{-1} \circ w' \circ \theta\right)(x),$$
$$w'(x') = \left(\theta \circ w \circ \theta^{-1}\right)(x'), \qquad (5.2)$$

where θ^{-1} denotes the inverse of θ.

In particular, if both θ and w are affine transformations of the Euclidean plane into itself, w' will still be an affine transformation, since the composition of affine transformations is still affine, and so are their inverses. Hence, if an IFS is given, consisting only of affine transformations, the IFS that results from an affine change of coordinates will still consist exclusively of affine transformations.

As an application, suppose we desire to produce an animation segment in which the spiral of Figure 2.7 spins around its center. If we apply a coordinate change θ consisting of a rotation by an angle δ around the center of the spiral, the resulting image will obviously be a rotated spiral. From a computational point of view, generation of such a rotated spiral amounts to calculating a new IFS by applying formula (5.2) to each transformation in the original one, and running the random iteration algorithm using the new IFS as input. Now, if the rotation angle δ is increased by small, equal amounts, and the procedure described above is repeated for each value of δ so determined, the resulting images, when viewed sequentially, will appear as a spiral spinning with uniform angular velocity around its center. If an angular velocity other than uniform is desired, it is enough to vary the rate at which the rotation angle δ is incremented. For instance, in order to obtain a uniformly accelerated spinning, δ could be incremented according to a quadratic function of the elapsed time.

By applying coordinate changes in a suitable manner, we can also easily zoom in and out of an image. Consider, once again, the dragon of Figure 2.3. The first transformation in its IFS encoding, listed on page 21, has its fixed point at $(0.75, 0.5)^T$. Suppose that our viewing window is given by the unit square, and that we want to produce an animation segment that, starting from an image of the dragon, moves such a fixed point towards the center of the window, while zooming into the image. Such a visual effect can be obtained by applying a coordinate change to the original IFS that dilates the original image, and translates the point $(0.75, 0.5)^T$ into the point $(0.5, 0.5)^T$, thus producing a new IFS, whose associated image can be used as the last frame of the required animation segment. The intermediate frames can be generated, as usual, by linear interpolation between corresponding transformations.

In particular, if we want the dilation factor to be equal to 10 along both coordinate axes, the coordinate change that we need is provided by:

$$\theta\left((x_1, x_2)^T\right) = \begin{pmatrix} 10 & 0 \\ 0 & 10 \end{pmatrix} \begin{pmatrix} x_1 \\ x_2 \end{pmatrix} + \begin{pmatrix} -7 \\ -4.5 \end{pmatrix}.$$

Applying formula (5.2), we obtain, after some algebraic manipulations, that the two transformations in the IFS encoding of the last frame are given by:

$$v_1\left((x_1, x_2)^T\right) = \begin{pmatrix} 0.5 & 0.5 \\ -0.5 & 0.5 \end{pmatrix} \begin{pmatrix} x_1 \\ x_2 \end{pmatrix} + \begin{pmatrix} 0 \\ 0.5 \end{pmatrix}$$

and

$$v_2\left((x_1, x_2)^T\right) = \begin{pmatrix} 0.5 & 0.5 \\ -0.5 & 0.5 \end{pmatrix} \begin{pmatrix} x_1 \\ x_2 \end{pmatrix} + \begin{pmatrix} -2.5 \\ -2 \end{pmatrix}.$$

Figure 5.6 contains some of the frames relative to this animation segment.

When, as a result of the change of coordinates, some points of the trajectory of the Markov chain generated through the random iteration algorithm fall outside the viewing window, the total number of points that are generated and plotted has to be increased accordingly. As a matter of fact, this way of operating is not very efficient. It has been pointed out by Marc Berger in a private communication that a better way of generating only part of the attractor of an IFS can be devised, based on the notion of addresses of points in the attractor. Such a notion is rigorously developed in Barnsley (1988), where it is shown that, as a consequence of the self-covering property, all points in the attractor A can be addressed (not necessarily

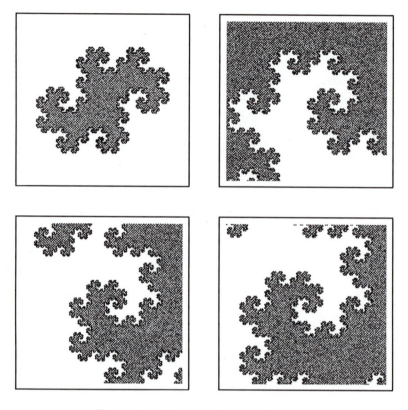

Figure 5.6. Dragon with levels of zoom-in.

uniquely) in terms of the sequences of transformations, applied to A, that lead to them.

For example, consider the case of the Sierpinski triangle of Figure 2.12, which is the attractor of the IFS given on page 20. As noted in Section 2.5, all images under w_1 of points in the attractor A fall in the lower left subtriangle $w_1(A)$ with vertices (F, E, C), and at least one of their addresses begins with 1. Similarly, all images under w_2 of points in $w_1(A)$ fall in the lower left subtriangle $w_2(w_1(A))$ contained in (D, B, E), and at least one of their addresses begins with 21. Continuing in this fashion, every point in the attractor can be assigned at least one address, consisting of an infinite sequence $\sigma_1\sigma_2\sigma_3\ldots$, where each σ_k belongs to $\{1, 2, 3\}$.

Returning to our application, suppose that our viewing window is contained in a region B of the attractor, consisting of points having addresses $\sigma_1\sigma_2\ldots\sigma_K\ldots$ that coincide up to the Kth position. It is then clear that, if a point generated by the random iteration algorithm is premultiplied by $w_{\sigma_1}\circ w_{\sigma_2}\circ\cdots\circ w_{\sigma_K}$, such a point will fall in B. In particular, if B is strictly contained in A and is not much bigger than the viewing window, the number of points to be discarded because they fall outside the window will be smaller than if the random iteration algorithm were applied in its original form.

Control can also be exerted on the resulting animation segment by employing IFSs with condensation, whose corresponding images can be easily generated by means of the mixing algorithm described in Section 3.5. As an example, consider the three frames of Figure 5.7 and Color Plate 8. The one at the bottom coincides with the image of Figure 2.11, and is obtained as the attractor of an IFS with condensation that, as we have seen on page 30, has the tallest pine tree on the left side as its condensation set, and is completed by the additional affine transformation that linearly attracts by a factor of 0.6 every point in the plane towards its fixed point $(1,0)^T$.

The first frame consists only of the pine tree on the left side. For the purpose of animation, it is expedient to also view this image as the attractor of an IFS with condensation, which has the tree as its condensation set, and is completed by the identity transformation. Notice, incidentally, that such an IFS is not hyperbolic. The intermediate frames are all obtained as attractors of IFSs with condensation. Each one of them has the tree on the left side as its condensation set, and is completed by an affine transformation whose fixed point is given by a convex combination of $(1,0)^T$ and the midpoint at the base of the tree. The weights of the combination are given by δ and $(1-\delta)$, respectively, with $0\leq\delta\leq 1$. In addition, such a transformation linearly attracts every point in the plane towards its fixed point by a factor of $(1-\delta)+(0.6)\delta$.

Hence, for $\delta=0$ and $\delta=1$, we obtain the first frame and the last frame of Figure 5.7, respectively, while the middle frame is obtained for $\delta=0.5$. The resulting animation segment consists of a row of pine trees growing from the left side of the screen towards the right.

We hope that the ideas and examples presented in this section have made it plausible, if not evident, that IFS techniques actually can be employed to make computer-animated motion pictures. While it is true that this area of application is still in its infancy, it

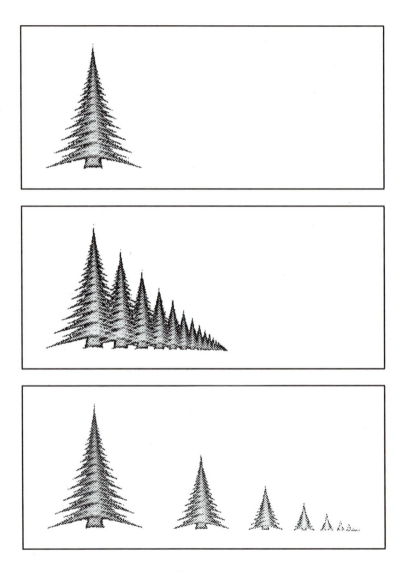

Figure 5.7. Growing row of pine trees.

appears to us that the savings in storage space that can be achieved through IFS encoding make it worthwhile pursuing research development in this direction, in spite of the inevitable difficulties that will be encountered along the way.

5.4. Graphics Equipment

We have made use of the equipment in the Graphics Laboratory of the Department of Statistics at Carnegie Mellon University to implement most of the ideas presented in this book, producing both still images and computer-animated videotapes. We would like to devote the present section to a brief, and not overly technical, description of the facility. [2]

The heart of the laboratory is a VAX 3200 color workstation. The other two major parts of the system are an analog film recorder, to make 35-mm slides, and a TV system, to make computer-animated videotapes. The schematic of the equipment presented in Figure 5.8 might help in understanding the following discussion.

The workstation generates a television signal, known as RGB (red, green, blue) video, and sends it to its color monitor and to the film recorder, which, essentially, can take photographs of the screen that do not suffer from the various distortions involved with taking a camera photograph of the screen. This is possible since a camera mounted on the instrument can take a photograph of an analog reproduction of the screen, which is generated inside the instrument itself.

The key to the computer-animated video system is a printed circuit board, which is inside the VAX workstation, and generates an RGB television signal (different, though, than the one used to drive the workstation monitor). Since the RGB video signal is different than the NTSC (National Television System Committee) video signal used in home television, and one would ultimately like to be able to produce videotapes that can be played on a standard VHS video cassette recorder, the RGB signal has to be converted into NTSC. In order to do so, the RGB signal generated by the board inside the VAX is sent to an NTSC color encoder that performs the conversion by rotating the RGB color cube and compressing the three color coordinates into one signal.

Another essential part of the TV system is a UMATIC video cas-

---

[2]While this description is based on the equipment available in 1990, we believe that it still provides an interesting and informative account. More powerful computing resources and more advanced recording devices (such as laser disk recorders) have more recently become available, but the basic steps involved in the production of computer animation have remained the same.

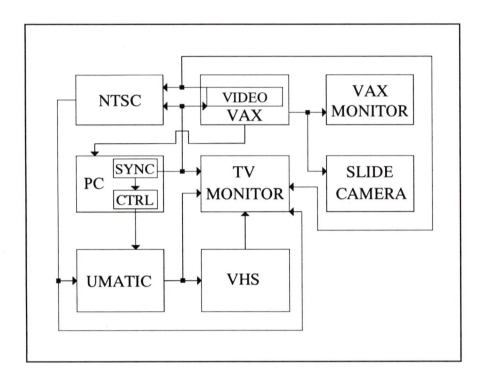

Figure 5.8. Schematic of the graphics equipment.

sette recorder, which uses (3/4)-inch videotapes, and has special editing capabilities. The UMATIC VCR can record a time code on one of two sound tracks on the videotape, and edit a single video frame at any specified location on the videotape. Animation is created by editing successive video frames on the tape, at the rate of 30 frames per second of animation. The control signals for editing are sent to the VCR by a controller which resides in an IBM personal computer.

The controller receives instructions from a program running on the VAX workstation. Such a program creates a digitized image, loads it into the video frame buffer on the board that generates the RGB video signal, and issues an editing instruction to the control unit in the PC. The control unit translates this instruction into control signals that are sent to the UMATIC VCR, which in turn edits a frame at the right point on the videotape. This cycle is repeated until all the frames comprising the animation sequence have been successively created and recorded on the videotape.

The equipment is completed by a VHS video cassette recorder, which is used to transfer animations created on UMATIC videotapes onto conventional VHS videotapes, and a Trinitron color monitor that can display both RGB video signal and NTSC video signal.

Bibliography

[1] Barnsley, M., and Demko, S. (1985). "Iterated Function Systems and the Global Construction of Fractals," *Proceedings of the Royal Society of London,* Ser. A, Vol. 399, 243–275.

[2] Barnsley, M., Ervin, V., Hardin, D. and Lancaster, J. (1986). "Solution of an Inverse Problem for Fractals and Other Sets," *Proceedings of the National Academy of Sciences, USA,* Vol. 83, 1975–1977.

[3] Barnsley, M. (1988). *Fractals Everywhere,* Academic Press, Boston.

[4] Barnsley, M., Berger, M., and Soner, H. (1988). "Mixing Markov Chains and Their Images," *Probability in the Engineering and Informational Sciences,* Vol. 2, 387–414.

[5] Barnsley, M., and Sloan, A. (1988). "A Better Way to Compress Images," *BYTE,* Vol. 13, January 1988, 215–223.

[6] Barnsley, M., and Hurd, L. (1993). *Fractal Image Compression,* AK Peters, Wellesley.

[7] Berger, M., and Amit, Y. (1987). "Products of Random Affine Maps," preprint, Weizmann Institute of Science, Rehovot, Israel.

[8] Berger, M. (1988). "Encoding Images through Transition Probabilities," *Proceedings of the Sixth International Conference on Mathematical Modeling, Mathl. Comput. Modeling,* Vol. 11, 575–577.

[9] Berger, M., and Soner, H. (1988). "Random Walks Generated by Affine Mappings," *Journal of Theoretical Probability,* Vol. 1, 239–254.

[10] Berger, M. (1989a). "Images Generated by Orbits of 2-D Markov Chains," *Chance,* Vol. 2, No. 2, 18–28.

[11] Berger, M. (1989b). "Random Affine Iterated Function Systems: Mixing and Encoding," Preprint Series, School of Mathematics, Georgia Institute of Technology, Atlanta.

[12] Berger, M. (1993). *An Introduction to Probability and Stochastic Processes,* Springer-Verlag, New York.

[13] David, H. (1981). *Order Statistics,* Second Edition, John Wiley & Sons, New York.

[14] Deans, S.R. (1983). *The Radon Transform and some of Its Applications,* Wiley Interscience, New York.

[15] Demko, S., Hodges, L., and Naylor, B. (1985). "Construction of Fractal Objects with Iterated Function Systems," *SIGGRAPH,* Vol. 19, No. 3, 271–278.

[16] Diaconis, P., and Shashahani, M. (1986). "Products of Random Matrices and Computer Image Generation," *Contemporary Mathematics,* Vol. 50, 173–182.

[17] Doob, J. (1953). *Stochastic Processes,* Wiley Interscience, New York.

[18] Edgar, G. (1990). *Measure, Topology, and Fractal Geometry,* Springer-Verlag, New-York.

[19] Elton, J. (1987). "An Ergodic Theorem for Iterated Maps," *Journal of Ergodic Theory and Dynamical Systems*, Vol. 7, 481–488.

[20] Falconer, K. (1990). *Fractal Geometry*, John Wiley & Sons, Chichester, Great Britain.

[21] Feller, W. (1968). *An Introduction to Probability Theory and Its Applications*, Third Edition, Vol. 1, John Wiley & Sons, New York.

[22] Foley, J., van Dam, A., Feiner, S., and Hughes, J. (1990). *Computer Graphics—Principles and Practice*, Second Edition, Addison Wesley, Reading.

[23] Hepting, D., Prusinkiewicz, P., and Saupe, D. (1991). "Rendering Methods for Iterated Function Systems," *Fractals in the Fundamental and Applied Sciences*, Peitgen, H.-O., Henriques, J., and Peneda, L. (eds.), North-Holland, Amsterdam.

[24] Hsu, C.S. (1987). *Cell-to-Cell Mapping (A Method of Global Analysis for Nonlinear Systems)*, Springer-Verlag, New York.

[25] Hutchinson, J. (1981). "Fractals and Self-Similarity," *Indiana University Journal of Mathematics*, Vol. 30, No. 5, 713–747.

[26] Jacquin, A. (1992). "Image Coding Based on a Fractal Theory of Iterated Contractive Image Transformations," *IEEE Transactions on Image Processing*, Vol. 1, 18–30.

[27] Kahan, W., Palmer, J., and Coonen, J. (1979). "A Proposed IEEE-CS Standard for Binary Floating Point Arithmetic," *Proceedings of the "Computer Science and Statistics: 12th Annual Symposium on the Interface,"* 32–36.

[28] Lax, P.D. (1971). "Approximation of Measure Preserving Transformations," *Communications on Pure and Applied Mathematics*, Vol. 24, 133–135.

[29] Lelewer, D.A., and Hirschberg, D.S. (1987). "Data Compression," *ACM Computing Surveys*, Vol. 19, No. 3, 261–296.

[30] Peitgen, H.-O., Jürgens, H., and Saupe, D. (1992). *Chaos and Fractals—New Frontiers of Science,* Springer-Verlag, New York.

[31] Peruggia, M. (1990). "Iterated Function Systems and the Propagation of Rounding Errors," Ph.D. Dissertation, Department of Statistics, Carnegie Mellon University, Pittsburgh.

[32] Ross, S. (1983). *Stochastic Processes,* John Wiley & Sons, New York.

[33] Rudin, W. (1976). *Principles of Mathematical Analysis,* Third Edition, McGraw-Hill, New York.

[34] Strang, S. (1980). *Linear Algebra and its Applications,* Second Edition, Academic Press, New York.

[35] Stevenson, D. (1981). "A Proposed Standard for Binary Floating-Point Arithmetic," *Computer,* Vol. 14, No. 3, 51–62.

[36] Zorpette, G. (1988). "Fractals: not just Another Pretty Picture," *IEEE Spectrum,* Vol. 25, October 1988, 29–31.

Index

Other Titles of Interest

————— from —————

A K PETERS, LTD

Barnsley/Hurd
Fractal Image Compression
ISBN 1-56881-000-8

Birmingham et al.
Automating the Design of Computer Systems
ISBN 0-86720-241-6

Boehm/Prautzsch
Geometric Concepts for Geometric Design
ISBN 1-56881-004-0

Cowan
Colour Principles for Computer Graphics
ISBN 1-56881-009-1

Hoschek/Lasser
Fundamentals of Computer Aided Geometric Design
ISBN 1-56881-007-5

Iterated Systems Inc.
Snapshots: True-Color Photo Images Using the Fractal Formatter
ISBN 0-86720-299-8

Klinker
A Physical Approach to Color Image Understanding
ISBN 1-56881-013-X

Parke/Waters
Computer Facial Animation
ISBN 1-56881-014-8

Whitman
Multiprocessor Methods for Computer Graphics Rendering
ISBN 0-86720-293-9

289 Linden Street
Wellesley, Massachusetts
(617) 235-2210
Fax (617) 235-2404